R and Data Mining

R and Data Mining
Examples and Case Studies

Yanchang Zhao

RDataMining.com

ELSEVIER

AMSTERDAM • BOSTON • HEIDELBERG • LONDON • NEW YORK
OXFORD • PARIS • SAN DIEGO • SAN FRANCISCO • SINGAPORE
SYDNEY • TOKYO
Academic Press is an imprint of Elsevier

Academic Press is an imprint of Elsevier
525 B Street, Suite 1900, San Diego, CA 92101-4495, USA
225 Wyman Street, Waltham, MA 02451, USA
32 Jamestown Road, London NW17BY, UK
Radarweg 29, PO Box 211, 1000 AE Amsterdam, The Netherlands

First edition 2013

Notice
No responsibility is assumed by the publisher for any injury and/or damage to persons or property as a matter of products liability, negligence or otherwise, or from any use or operation of any methods, products, instructions or ideas contained in the material herein. Because of rapid advances in the medical sciences, in particular, independent verification of diagnoses and drug dosages should be made.

Library of Congress Cataloging-in-Publication Data
Application submitted

British Library Cataloguing-in-Publication Data
A catalogue record for this book is available from the British Library

ISBN: 978-0-123-96963-7

For information on all Academic Press publications visit
our website at store.elsevier.com

To Yanbo, Michael and Lucas for your love and encouragement

Contents

List of Figures   xi

List of Abbreviations   xv

1   Introduction   **1**
    1.1   Data Mining   1
    1.2   R   2
    1.3   Datasets   2
       1.3.1   The Iris Dataset   2
       1.3.2   The Bodyfat Dataset   3

2   Data Import and Export   **5**
    2.1   Save and Load R Data   5
    2.2   Import from and Export to .CSV Files   5
    2.3   Import Data from SAS   6
    2.4   Import/Export via ODBC   8
       2.4.1   Read from Databases   8
       2.4.2   Output to and Input from EXCEL Files   9

3   Data Exploration   **11**
    3.1   Have a Look at Data   11
    3.2   Explore Individual Variables   13
    3.3   Explore Multiple Variables   16
    3.4   More Explorations   20
    3.5   Save Charts into Files   25

4   Decision Trees and Random Forest   **27**
    4.1   Decision Trees with Package *party*   27
    4.2   Decision Trees with Package *rpart*   31
    4.3   Random Forest   36

5   Regression   **41**
    5.1   Linear Regression   41
    5.2   Logistic Regression   47
    5.3   Generalized Linear Regression   48
    5.4   Non-Linear Regression   50

6   Clustering   **51**
    6.1   The k-Means Clustering   51
    6.2   The k-Medoids Clustering   53

6.3	Hierarchical Clustering	56
6.4	Density-Based Clustering	57

7 Outlier Detection — **63**
7.1	Univariate Outlier Detection	63
7.2	Outlier Detection with LOF	66
7.3	Outlier Detection by Clustering	70
7.4	Outlier Detection from Time Series	72
7.5	Discussions	73

8 Time Series Analysis and Mining — **75**
8.1	Time Series Data in R	75
8.2	Time Series Decomposition	76
8.3	Time Series Forecasting	78
8.4	Time Series Clustering	78
	8.4.1 Dynamic Time Warping	79
	8.4.2 Synthetic Control Chart Time Series Data	79
	8.4.3 Hierarchical Clustering with Euclidean Distance	80
	8.4.4 Hierarchical Clustering with DTW Distance	82
8.5	Time Series Classification	83
	8.5.1 Classification with Original Data	83
	8.5.2 Classification with Extracted Features	84
	8.5.3 k-NN Classification	86
8.6	Discussions	87
8.7	Further Readings	87

9 Association Rules — **89**
9.1	Basics of Association Rules	89
9.2	The Titanic Dataset	90
9.3	Association Rule Mining	92
9.4	Removing Redundancy	96
9.5	Interpreting Rules	98
9.6	Visualizing Association Rules	99
9.7	Discussions and Further Readings	103

10 Text Mining — **105**
10.1	Retrieving Text from Twitter	105
10.2	Transforming Text	106
10.3	Stemming Words	108
10.4	Building a Term-Document Matrix	110
10.5	Frequent Terms and Associations	111
10.6	Word Cloud	113
10.7	Clustering Words	114
10.8	Clustering Tweets	116

	10.8.1	Clustering Tweets with the k-Means Algorithm	**116**
	10.8.2	Clustering Tweets with the k-Medoids Algorithm	**118**
10.9	Packages, Further Readings, and Discussions		**121**

11 Social Network Analysis **123**
11.1	Network of Terms	**123**
11.2	Network of Tweets	**127**
11.3	Two-Mode Network	**132**
11.4	Discussions and Further Readings	**136**

12 Case Study I: Analysis and Forecasting of House Price Indices **137**
12.1	Importing HPI Data	**137**
12.2	Exploration of HPI Data	**138**
12.3	Trend and Seasonal Components of HPI	**145**
12.4	HPI Forecasting	**147**
12.5	The Estimated Price of a Property	**149**
12.6	Discussion	**149**

13 Case Study II: Customer Response Prediction and Profit Optimization **151**
13.1	Introduction	**151**
13.2	The Data of KDD Cup 1998	**151**
13.3	Data Exploration	**160**
13.4	Training Decision Trees	**166**
13.5	Model Evaluation	**170**
13.6	Selecting the Best Tree	**173**
13.7	Scoring	**176**
13.8	Discussions and Conclusions	**179**

14 Case Study III: Predictive Modeling of Big Data with Limited Memory **181**
14.1	Introduction	**181**
14.2	Methodology	**182**
14.3	Data and Variables	**182**
14.4	Random Forest	**183**
14.5	Memory Issue	**185**
14.6	Train Models on Sample Data	**186**
14.7	Build Models with Selected Variables	**188**
14.8	Scoring	**194**
14.9	Print Rules	**201**
	14.9.1 Print Rules in Text	**201**
	14.9.2 Print Rules for Scoring with SAS	**205**
14.10	Conclusions and Discussion	**211**

15   Online Resources   **213**
   15.1   R Reference Cards   213
   15.2   R   213
   15.3   Data Mining   214
   15.4   Data Mining with R   216
   15.5   Classification/Prediction with R   216
   15.6   Time Series Analysis with R   216
   15.7   Association Rule Mining with R   216
   15.8   Spatial Data Analysis with R   217
   15.9   Text Mining with R   217
   15.10 Social Network Analysis with R   217
   15.11 Data Cleansing and Transformation with R   218
   15.12 Big Data and Parallel Computing with R   218

R Reference Card for Data Mining   **221**

Bibliography   **225**

General Index   **229**

Package Index   **231**

Function Index   **233**

List of Figures

3.1	Histogram	15
3.2	Density	15
3.3	Pie Chart	16
3.4	Bar Chart	16
3.5	Boxplot	18
3.6	Scatter Plot	18
3.7	Scatter Plot with Jitter	19
3.8	A Matrix of Scatter Plots	19
3.9	3D Scatter Plot	20
3.10	Heat Map	21
3.11	Level Plot	22
3.12	Contour	22
3.13	3D Surface	23
3.14	Parallel Coordinates	23
3.15	Parallel Coordinates with Package *lattice*	24
3.16	Scatter Plot with Package *ggplot2*	24
4.1	Decision Tree	29
4.2	Decision Tree (Simple Style)	30
4.3	Decision Tree with Package *rpart*	34
4.4	Selected Decision Tree	35
4.5	Prediction Result	36
4.6	Error Rate of Random Forest	38
4.7	Variable Importance	39
4.8	Margin of Predictions	40
5.1	Australian CPIs in Year 2008 to 2010	42
5.2	Prediction with Linear Regression Model	45
5.3	A 3D Plot of the Fitted Model	46
5.4	Prediction of CPIs in 2011 with Linear Regression Model	47
5.5	Prediction with Generalized Linear Regression Model	50
6.1	Results of k-Means Clustering	53
6.2	Clustering with the k-medoids Algorithm—I	54
6.3	Clustering with the k-medoids Algorithm—II	55
6.4	Cluster Dendrogram	56
6.5	Density-Based Clustering—I	58
6.6	Density-Based Clustering—II	59
6.7	Density-Based Clustering—III	59
6.8	Prediction with Clustering Model	60

7.1	Univariate Outlier Detection with Boxplot	64
7.2	Outlier Detection—I	65
7.3	Outlier Detection—II	66
7.4	Density of Outlier Factors	67
7.5	Outliers in a Biplot of First Two Principal Components	68
7.6	Outliers in a Matrix of Scatter Plots	69
7.7	Outliers with k-Means Clustering	71
7.8	Outliers in Time Series Data	73

8.1	A Time Series of AirPassengers	76
8.2	Seasonal Component	77
8.3	Time Series Decomposition	77
8.4	Time Series Forecast	78
8.5	Alignment with Dynamic Time Warping	79
8.6	Six Classes in Synthetic Control Chart Time Series	80
8.7	Hierarchical Clustering with Euclidean Distance	81
8.8	Hierarchical Clustering with DTW Distance	82
8.9	Decision Tree	84
8.10	Decision Tree with DWT	86

9.1	A Scatter Plot of Association Rules	100
9.2	A Balloon Plot of Association Rules	100
9.3	A Graph of Association Rules	101
9.4	A Graph of Items	102
9.5	A Parallel Coordinates Plot of Association Rules	102

10.1	Frequent Terms	112
10.2	Word Cloud	114
10.3	Clustering of Words	115
10.4	Clusters of Tweets	120

11.1	A Network of Terms—I	125
11.2	A Network of Terms—II	126
11.3	Distribution of Degree	128
11.4	A Network of Tweets—I	129
11.5	A Network of Tweets—II	130
11.6	A Network of Tweets—III	131
11.7	A Two-Mode Network of Terms and Tweets—I	133
11.8	A Two-Mode Network of Terms and Tweets—II	135

12.1	HPIs in Canberra from Jan. 1990 to Jan. 2011	139
12.2	Monthly Increase of HPI	140
12.3	Monthly Increase Rate of HPI	141
12.4	A Bar Chart of Monthly HPI Increase Rate	142
12.5	Number of Months with Increased HPI	143
12.6	Yearly Average Increase Rates of HPI	143
12.7	Monthly Average Increase Rates of HPI	144
12.8	Distribution of HPI Increase Rate	144
12.9	Distribution of HPI Increase Rate per Year	145

12.10	Distribution of HPI Increase Rate per Month	145
12.11	Decomposition of HPI Data	146
12.12	Seasonal Components of HPI Data	146
12.13	HPI Forecasting—I	148
12.14	HPI Forecasting—II	149
13.1	A Data Mining Process	152
13.2	Distribution of Response	156
13.3	Box Plot of Donation Amount	156
13.4	Barplot of Donation Amount	157
13.5	Histograms of Numeric Variables	161
13.6	Boxplot of HIT	162
13.7	Distribution of Donation in Various Age Groups	163
13.8	Distribution of Donation in Various Age Groups	164
13.9	Scatter Plot	165
13.10	Mosaic Plots of Categorical Variables	166
13.11	A Decision Tree	169
13.12	Total Donation Collected (1000—400—4—10)	171
13.13	Total Donation Collected (9 runs)	172
13.14	Average Result of Nine Runs	173
13.15	Comparison of Different Parameter Settings—I	175
13.16	Comparison of Different Parameter Settings—II	175
13.17	Validation Result	178
14.1	Decision Tree	191
14.2	Test Result—I	192
14.3	Test Result—II	193
14.4	Test Result—III	194
14.5	Distribution of Scores	200

List of Abbreviations

ARIMA	Autoregressive integrated moving average
ARMA	Autoregressive moving average
AVF	Attribute value frequency
CLARA	Clustering for large applications
CRISP-DM	Cross industry standard process for data mining
DBSCAN	Density-based spatial clustering of applications with noise
DTW	Dynamic time warping
DWT	Discrete wavelet transform
GLM	Generalized linear model
IQR	Interquartile range, i.e., the range between the first and third quartiles
LOF	Local outlier factor
PAM	Partitioning around medoids
PCA	Principal component analysis
STL	Seasonal-trend decomposition based on Loess
TF-IDF	Term frequency-inverse document frequency

1 Introduction

This book introduces into using R for data mining. It presents many examples of various data mining functionalities in R and three case studies of real-world applications. The supposed audience of this book are postgraduate students, researchers, and data miners who are interested in using R to do their data mining research and projects. We assume that readers already have a basic idea of data mining and also have some basic experience with R. We hope that this book will encourage more and more people to use R to do data mining work in their research and applications.

This chapter introduces basic concepts and techniques for data mining, including a data mining process and popular data mining techniques. It also presents R and its packages, functions, and task views for data mining. At last, some datasets used in this book are described.

1.1 Data Mining

Data mining is the process to discover interesting knowledge from large amounts of data (Han and Kamber, 2000). It is an interdisciplinary field with contributions from many areas, such as statistics, machine learning, information retrieval, pattern recognition, and bioinformatics. Data mining is widely used in many domains, such as retail, finance, telecommunication, and social media.

The main techniques for data mining include classification and prediction, clustering, outlier detection, association rules, sequence analysis, time series analysis, and text mining, and also some new techniques such as social network analysis and sentiment analysis. Detailed introduction of data mining techniques can be found in text books on data mining (Han and Kamber, 2000; Hand et al., 2001; Witten and Frank, 2005). In real-world applications, a data mining process can be broken into six major phases: business understanding, data understanding, data preparation, modeling, evaluation, and deployment, as defined by the CRISP-DM (Cross Industry Standard Process for Data Mining).[1] This book focuses on the modeling phase, with data exploration and model evaluation involved in some chapters. Readers who want more information on data mining are referred to online resources in Chapter 15.

---

[1] http://www.crisp-dm.org/.

R and Data Mining. DOI: http://dx.doi.org/10.1016/B978-0-12-396963-7.00001-5

RDM

1.2   R

R[2] (R Development Core Team, 2012) is a free software environment for statistical computing and graphics. It provides a wide variety of statistical and graphical techniques. R can be extended easily via packages. There are around 4000 packages available in the CRAN package repository,[3] as on August 1, 2012. More details about R are available in *An Introduction to R*[4] (Venables et al., 2012) and *R Language Definition*[5] (R Development Core Team, 2010b) at the CRAN website. R is widely used in both academia and industry.

To help users to find out which R packages to use, the CRAN Task Views[6] are a good guidance. They provide collections of packages for different tasks. Some task views related to data mining are:

- Machine Learning and Statistical Learning;

- Cluster Analysis and Finite Mixture Models;

- Time Series Analysis;

- Multivariate Statistics; and

- Analysis of Spatial Data.

Another guide to R for data mining is an *R Reference Card for Data Mining* (see p. 221), which provides a comprehensive indexing of R packages and functions for data mining, categorized by their functionalities. Its latest version is available at http://www.rdatamining.com/docs.

Readers who want more information on R are referred to online resources in Chapter 15.

1.3   Datasets

The datasets used in this book are briefly described in this section.

1.3.1   The Iris Dataset

The iris dataset has been used for classification in many research publications. It consists of 50 samples from each of three classes of iris flowers (Frank and Asuncion, 2010). One class is linearly separable from the other two, while the latter are not linearly separable from each other. There are five attributes in the dataset:

---

[2] http://www.r-project.org/.
[3] http://cran.r-project.org/.
[4] http://cran.r-project.org/doc/manuals/R-intro.pdf.
[5] http://cran.r-project.org/doc/manuals/R-lang.pdf.
[6] http://cran.r-project.org/web/views/.

- sepal length in cm,

- sepal width in cm,

- petal length in cm,

- petal width in cm, and

- class: Iris Setosa, Iris Versicolour, and Iris Virginica.

```
> str(iris)

'data.frame':  150 obs.of 5 variables:

$ Sepal.Length: num 5.1 4.9 4.7 4.6 5 5.4 4.6 5 4.4 4.9 …

$ Sepal.Width: num 3.5 3 3.2 3.1 3.6 3.9 3.4 3.4 2.9 3.1 …

$ Petal.Length: num 1.4 1.4 1.3 1.5 1.4 1.7 1.4 1.5 1.4 1.5 …

$ Petal.Width: num 0.2 0.2 0.2 0.2 0.2 0.4 0.3 0.2 0.2 0.1 …

$ Species: Factor w/ 3 levels "setosa","versicolor",..: 1 1 1 1 1
  1 1 1 1 1 …
```

1.3.2   The Bodyfat Dataset

Bodyfat is a dataset available in package *mboost* (Hothorn et al., 2012). It has 71 rows, and each row contains information of one person. It contains the following 10 numeric columns:

- age: age in years.

- DEXfat: body fat measured by DXA, response variable.

- waistcirc: waist circumference.

- hipcirc: hip circumference.

- elbowbreadth: breadth of the elbow.

- kneebreadth: breadth of the knee.

- anthro3a: sum of logarithm of three anthropometric measurements.

- anthro3b: sum of logarithm of three anthropometric measurements.

- anthro3c: sum of logarithm of three anthropometric measurements.

- anthro4: sum of logarithm of three anthropometric measurements.

The value of DEXfat is to be predicted by the other variables:

```
> data("bodyfat", package = "mboost")

> str(bodyfat)

'data.frame':      71 obs. of 10 variables:

$ age: num 57 65 59 58 60 61 56 60 58 62 …

$ DEXfat: num 41.7 43.3 35.4 22.8 36.4 …

$ waistcirc: num 100 99.5 96 72 89.5 83.5 81 89 80 79 …

$ hipcirc: num 112 116.5 108.5 96.5 100.5 …

$ elbowbreadth: num 7.1 6.5 6.2 6.1 7.1 6.5 6.9 6.2 6.4 7 …

$ kneebreadth: num 9.4 8.9 8.9 9.2 10 8.8 8.9 8.5 8.8 8.8 …

$ anthro3a: num 4.42 4.63 4.12 4.03 4.24 3.55 4.14 4.04 3.91 3.66
  …

$ anthro3b: num 4.95 5.01 4.74 4.48 4.68 4.06 4.52 4.7 4.32 4.21
  …

$ anthro3c: num 4.5 4.48 4.6 3.91 4.15 3.64 4.31 4.47 3.47 3.6 …

$ anthro4: num 6.13 6.37 5.82 5.66 5.91 5.14 5.69 5.7 5.49 5.25 …
```

2 Data Import and Export

This chapter shows how to import foreign data into R and export R objects to other formats. At first, examples are given to demonstrate saving R objects to and loading them from .Rdata files. After that, it demonstrates importing data from and exporting data to .csv files, SAS databases, ODBC databases, and EXCEL files. For more details on data import and export, please refer to *R Data Import/Export*[1] (R Development Core Team, 2010a).

2.1 Save and Load R Data

Data in R can be saved as .Rdata files with function save( ). After that, they can then be loaded into R with load( ). In the code below, function rm( ) removes object a from R:

```
> a <- 1:10

> save(a, file="./data/dumData.Rdata")

> rm(a)

> load("./data/dumData.Rdata")

> print(a)

[1]  1  2  3  4  5  6  7  8  9 10
```

2.2 Import from and Export to .csv Files

The example below creates a dataframe df1 and saves it as a .csv file with write.csv( ). And then, the dataframe is loaded from file to df2 with read.csv( ):

```
> var1 <- 1:5

> var2 <- (1:5) / 10
```

---

[1] http://cran.r-project.org/doc/manuals/R-data.pdf.

R and Data Mining. DOI: http://dx.doi.org/10.1016/B978-0-12-396963-7.00002-7

```
> var3 <- c("R", "and", "Data Mining", "Examples", "Case
  Studies")

> df1 <- data.frame(var1, var2, var3)

> names(df1) <- c("VariableInt", "VariableReal", "VariableChar")

> write.csv(df1, "./data/dummmyData.csv", row.names = FALSE)

> df2 <- read.csv("./data/dummmyData.csv")

> print(df2)

    VariableInt   VariableReal   VariableChar

1   1             0.1            R

2   2             0.2            and

3   3             0.3            Data Mining

4   4             0.4            Examples

5   5             0.5            Case Studies
```

2.3   Import Data from SAS

Package *foreign* (R-core, 2012) provides function read.ssd() for importing SAS datasets (.sas7bdat files) into R. However, the following points are essential to make importing successful:

- SAS must be available on your computer, and read.ssd() will call SAS to read SAS datasets and import them into R.

- The file name of a SAS dataset has to be no longer than eight characters. Otherwise, the importing would fail. There is no such limit when importing from a .csv file.

- During importing, variable names longer than eight characters are truncated to eight characters, which often makes it difficult to know the meanings of variables. One way to get around this issue is to import variable names separately from a .csv file, which keeps full names of variables.

  An empty .csv file with variable names can be generated with the following method:

1. Create an empty SAS table dumVariables from dumData as follows:

```
data work.dumVariables;

  set work.dumData(obs=0);

run;
```

2. Export table `dumVariables` as a .CSV file.

The example below demonstrates importing data from a SAS dataset. Assume that there is a SAS data file `dumData.sas7bdat` and a .CSV file `dumVariables.csv` in folder `"Current working directory/data"`:

```
> library(foreign) # for importing SAS data

> # the path of SAS on your computer

> sashome <- "C:/Program Files/SAS/SASFoundation/9.2"

> filepath <- "./data"

> # filename should be no more than 8 characters, without
  extension

> fileName <- "dumData"

> # read data from a SAS dataset

> a <- read.ssd(file.path(filepath), fileName,
  sascmd=file.path(sashome, "sas.exe"))

> print(a)
```

```
      VARIABLE   VARIABL2   VARIABL3
1     1          0.1        R

2     2          0.2        and

3     3          0.3        Data Mining

4     4          0.4        Examples

5     5          0.5        Case Studies
```

Note that the variable names above are truncated. The full names can be imported from a .CSV file with the following code:

```
> # read variable names from a .CSV file

> variableFileName <- "dumVariables.csv"

> myNames <- read.csv(paste(filepath, variableFileName, sep="/"))

> names(a) <- names(myNames)

> print(a)
```

	VariableInt	VariableReal	VariableChar
1	1	0.1	R
2	2	0.2	and
3	3	0.3	Data Mining
4	4	0.4	Examples
5	5	0.5	Case Studies

Although one can export a SAS dataset to a .csv file and then import data from it, there are problems when there are special formats in the data, such as a value of "$100,000" for a numeric variable. In this case, it would be better to import from a .sas7bdat file. However, variable names may need to be imported into R separately as above.

Another way to import data from a SAS dataset is to use function read.xport() to read a file in SAS Transport (XPORT) format.

2.4   Import/Export via ODBC

Package *RODBC* provides connection to ODBC databases (Ripley and from 1999 to Oct 2002 Michael Lapsley, 2012).

2.4.1   Read from Databases

Below is an example of reading from an ODBC database. Function odbcConnect() sets up a connection to database, sqlQuery() sends an SQL query to the database, and odbcClose() closes the connection:

```
> library(RODBC)

> connection <- odbcConnect(dsn="servername",uid="userid",
  pwd="******")

> query <- "SELECT * FROM lib.table WHERE ..."

> # or read query from file

> # query <- readChar("data/myQuery.sql", nchars=99999)

> myData <- sqlQuery(connection, query, errors=TRUE)

> odbcClose(connection)
```

There are also sqlSave() and sqlUpdate() for writing or updating a table in an ODBC database.

2.4.2   Output to and Input from EXCEL Files

An example of writing data to and reading data from EXCEL files is shown below:

```
> library(RODBC)

> filename <- "data/dummmyData.xls"

> xlsFile <- odbcConnectExcel(filename, readOnly = FALSE)

> sqlSave(xlsFile, a, rownames = FALSE)

> b <- sqlFetch(xlsFile, "a")

> odbcClose(xlsFile)
```

Note that there might be a limit of 65,536 rows to write to an EXCEL file.

3 Data Exploration

This chapter shows examples on data exploration with R. It starts with inspecting the dimensionality, structure, and data of an R object, followed by basic statistics and various charts like pie charts and histograms. Exploration of multiple variables is then demonstrated, including grouped distribution, grouped boxplots, scattered plot, and pairs plot. After that, examples are given on level plot, contour plot, and 3D plot. It also shows how to save charts into files of various formats.

3.1 Have a Look at Data

The `iris` data is used in this chapter for demonstration of data exploration with R. See Section 1.3.1 for details of the `iris` data.

We first check the size and structure of data. The dimension and names of data can be obtained respectively with `dim()` and `names()`. Functions `str()` and `attributes()` return the structure and attributes of data.

```
> dim(iris)

[1] 150 5

> names(iris)

[1] "Sepal.Length" "Sepal.Width" "Petal.Length" "Petal.Width"
  "Species"

> str(iris)

'data.frame':    150 obs. of 5 variables:

$ Sepal.Length: num 5.1 4.9 4.7 4.6 5 5.4 4.6 5 4.4 4.9 …

$ Sepal.Width: num 3.5 3 3.2 3.1 3.6 3.9 3.4 3.4 2.9 3.1 …

$ Petal.Length: num 1.4 1.4 1.3 1.5 1.4 1.7 1.4 1.5 1.4 1.5 …

$ Petal.Width: num 0.2 0.2 0.2 0.2 0.2 0.4 0.3 0.2 0.2 0.1 …

$ Species     : Factor w/ 3 levels "setosa","versicolor",..: 1 1 1
  1 1 1 1 1 …
```

R and Data Mining. DOI: http://dx.doi.org/10.1016/B978-0-12-396963-7.00003-9

```
> attributes(iris)
```

```
$names
```

```
[1] "Sepal.Length" "Sepal.Width" "Petal.Length" "Petal.Width"
  "Species"
```

```
$row.names
```

[1]	1	2	3	4	5	6	7	8	9	10	11	12	13	14	15	16	17	18
[19]	19	20	21	22	23	24	25	26	27	28	29	30	31	32	33	34	35	36
[37]	37	38	39	40	41	42	43	44	45	46	47	48	49	50	51	52	53	54
[55]	55	56	57	58	59	60	61	62	63	64	65	66	67	68	69	70	71	72
[73]	73	74	75	76	77	78	79	80	81	82	83	84	85	86	87	88	89	90
[91]	91	92	93	94	95	96	97	98	99	100	101	102	103	104	105	106	107	108
[109]	109	110	111	112	113	114	115	116	117	118	119	120	121	122	123	124	125	126
[127]	127	128	129	130	131	132	133	134	135	136	137	138	139	140	141	142	143	144
[145]	145	146	147	148	149	150												

```
$class
```

```
[1] "data.frame"
```

Next, we have a look at the first five rows of data. The first or last rows of data can be retrieved with head() or tail().

```
> iris[1:5,]
```

	Sepal.Length	Sepal.Width	Petal.Length	Petal.Width	Species
1	5.1	3.5	1.4	0.2	setosa
2	4.9	3.0	1.4	0.2	setosa
3	4.7	3.2	1.3	0.2	setosa
4	4.6	3.1	1.5	0.2	setosa
5	5.0	3.6	1.4	0.2	setosa

```
> head(iris)
```

	Sepal.Length	Sepal.Width	Petal.Length	Petal.Width	Species
1	5.1	3.5	1.4	0.2	setosa
2	4.9	3.0	1.4	0.2	setosa
3	4.7	3.2	1.3	0.2	setosa
4	4.6	3.1	1.5	0.2	setosa
5	5.0	3.6	1.4	0.2	setosa
6	5.4	3.9	1.7	0.4	setosa

```
> tail(iris)
```

	Sepal.Length	Sepal.Width	Petal.Length	Petal.Width	Species
145	6.7	3.3	5.7	2.5	virginica
146	6.7	3.0	5.2	2.3	virginica
147	6.3	2.5	5.0	1.9	virginica
148	6.5	3.0	5.2	2.0	virginica
149	6.2	3.4	5.4	2.3	virginica
150	5.9	3.0	5.1	1.8	virginica

We can also retrieve the values of a single column. For example, the first 10 values of Sepal.Length can be fetched with either of the codes below.

```
> iris[1:10, "Sepal.Length"]
[1] 5.1 4.9 4.7 4.6 5.0 5.4 4.6 5.0 4.4 4.9
> iris$Sepal.Length[1:10]
[1] 5.1 4.9 4.7 4.6 5.0 5.4 4.6 5.0 4.4 4.9
```

3.2 Explore Individual Variables

Distribution of every numeric variable can be checked with function summary(), which returns the minimum, maximum, mean, median, and the first (25%) and third (75%) quartiles. For factors (or categorical variables), it shows the frequency of every level.

```
> summary(iris)
```

Sepal.Length	Sepal.Width	Petal.Length	Petal.Width	Species
Min.:4.300	Min.:2.000	Min.:1.000	Min.:0.100	setosa:50
1st Qu.:5.100	1st Qu.:2.800	1st Qu.:1.600	1st Qu.:0.300	versicolor:50
Median:5.800	Median:3.000	Median:4.350	Median:1.300	virginica:50
Mean:5.843	Mean:3.057	Mean:3.758	Mean:1.199	
3rd Qu.:6.400	3rd Qu.:3.300	3rd Qu.:5.100	3rd Qu.:1.800	
Max.:7.900	Max.:4.400	Max.:6.900	Max.:2.500	

The mean, median, and range can also be obtained with functions with mean(), median(), and range(). Quartiles and percentiles are supported by function quantile() as below.

```
> quantile(iris$Sepal.Length)
```

```
 0%   25%   50%   75%  100%
4.3   5.1   5.8   6.4   7.9
```

```
> quantile(iris$Sepal.Length, c(.1,.3,.65))
```

```
 10%   30%   65%
4.80  5.27  6.20
```

Then we check the variance of Sepal.Length with var() and its distribution with histogram and density using functions hist() and density() (see Figures 3.1 and 3.2).

```
> var(iris$Sepal.Length)
[1] 0.6856935
> hist(iris$Sepal.Length)
```

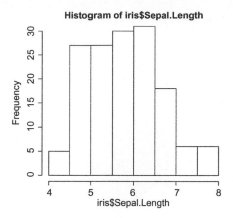

Figure 3.1 Histogram.

```
> plot(density(iris$Sepal.Length))
```

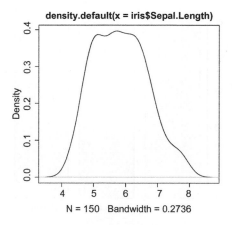

Figure 3.2 Density.

The frequency of factors can be calculated with function `table()` and then plotted as a pie chart with `pie()` or a bar chart with `barplot()` (see Figures 3.3 and 3.4).

```
> table(iris$Species)
```

```
    setosa    versicolor    virginica

    50           50             50
```

```
> pie(table(iris$Species))
```

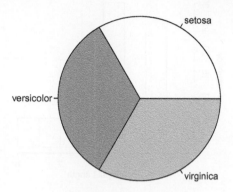

Figure 3.3 Pie chart.

```
> barplot(table(iris$Species))
```

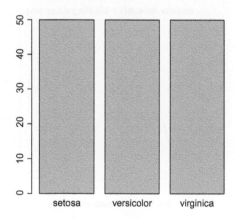

Figure 3.4 Bar chart.

3.3 Explore Multiple Variables

After checking the distributions of individual variables, we then investigate the relation-
ships between two variables. Below we calculate covariance and correlation between
variables with cov() and cor().

```
> cov(iris$Sepal.Length, iris$Petal.Length)

[1] 1.274315

> cov(iris[,1:4])
```

	Sepal.Length	Sepal.Width	Petal.Length	Petal.Width
Sepal.Length	0.6856935	−0.0424340	1.2743154	0.5162707
Sepal.Width	−0.0424340	0.1899794	−0.3296564	−0.1216394
Petal.Length	1.2743154	−0.3296564	3.1162779	1.2956094
Petal.Width	0.5162707	−0.1216394	1.2956094	0.5810063

```
> cor(iris$Sepal.Length, iris$Petal.Length)
```

[1] 0.8717538

```
> cor(iris[,1:4])
```

	Sepal.Length	Sepal.Width	Petal.Length	Petal.Width
Sepal.Length	1.0000000	−0.1175698	0.8717538	0.8179411
Sepal.Width	−0.1175698	1.0000000	−0.4284401	−0.3661259
Petal.Length	0.8717538	−0.4284401	1.0000000	0.9628654
Petal.Width	0.8179411	−0.3661259	0.9628654	1.0000000

Next, we compute the stats of Sepal.Length of every Species with aggregate().

```
> aggregate(Sepal.Length~Species, summary, data=iris)
```

	Species	Sepal.Length.Min.	Sepal.Length.1st Qu.	Sepal.Length.Median
1	setosa	4.300	4.800	5.000
2	versicolor	4.900	5.600	5.900
3	virginica	4.900	6.225	6.500

	Sepal.Length.Mean	Sepal.Length.3rd Qu.	Sepal.Length.Max.
1	5.006	5.200	5.800
2	5.936	6.300	7.000
3	6.588	6.900	7.900

We then use function boxplot() to plot a box plot, also known as box-and-whisker plot, to show the median, first and third quartiles of a distribution (i.e. the 50%, 25%, and 75% points in cumulative distribution), and outliers. The bar in the middle is the median. The box shows the interquartile range (IQR), which is the range between the 75% and 25% observation (see Figure 3.5).

```
> boxplot(Sepal.Length~Species, data=iris)
```

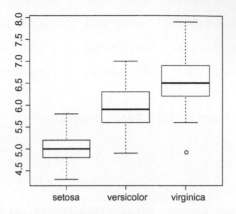

Figure 3.5 Boxplot.

A scatter plot can be drawn for two numeric variables with `plot()` as below. Using function `with()`, we do not need to add "`iris$`" before variable names. In the code below, the colors (`col`) and symbols (`pch`) of points are set to `Species` (see Figure 3.6).

```
> with(iris, plot(Sepal.Length, Sepal.Width, col=Species,
  pch=as.numeric(Species)))
```

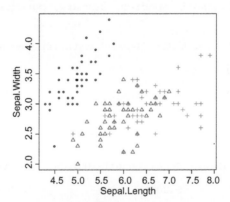

Figure 3.6 Scatter plot.

When there are many points, some of them may overlap. We can use `jitter()` to add a small amount of noise to the data before plotting (see Figure 3.7).

```
> plot(jitter(iris$Sepal.Length), jitter(iris$Sepal.Width))
```

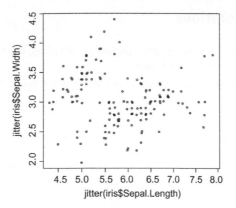

Figure 3.7 Scatter plot with jitter.

A matrix of scatter plots can be produced with function `pairs()` (see Figure 3.8).

```
> pairs(iris)
```

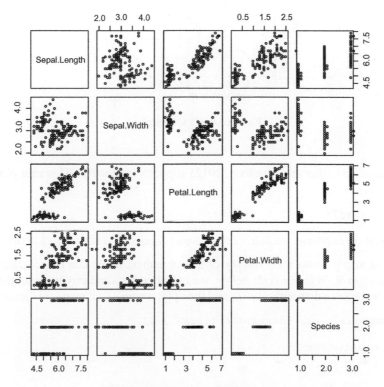

Figure 3.8 A matrix of scatter plots.

3.4   More Explorations

This section presents some fancy graphs, including 3D plots, level plots, contour plots, interactive plots, and parallel coordinates.

A 3D scatter plot can be produced with package *scatterplot3d* (Ligges and Mächler, 2003) (see Figure 3.9).

```
> library(scatterplot3d)

> scatterplot3d(iris$Petal.Width, iris$Sepal.Length,
  iris$Sepal.Width)
```

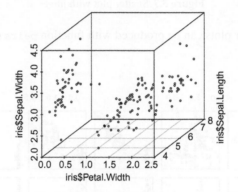

Figure 3.9 3D scatter plot.

Package *rgl* (Adler and Murdoch, 2012) supports interactive 3D scatter plot with `plot3d()`.

```
> library(rgl)

> plot3d(iris$Petal.Width, iris$Sepal.Length, iris$Sepal.Width)
```

A heat map presents a 2D display of a data matrix, which can be generated with `heatmap()` in R. With the code below, we calculate the similarity between different flowers in the `iris` data with `dist()` and then plot it with a heat map (see Figure 3.10).

```
> distMatrix <- as.matrix(dist(iris[,1:4]))

> heatmap(distMatrix)
```

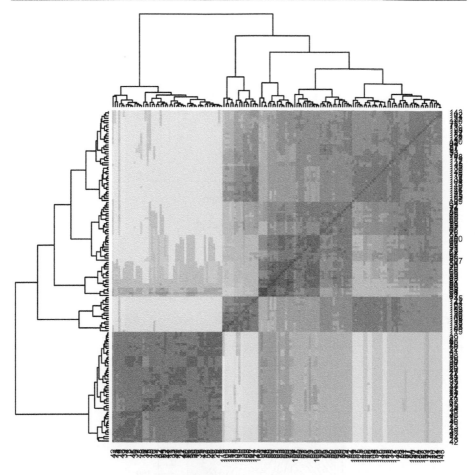

Figure 3.10 Heat map.

A level plot can be produced with function `levelplot()` in package *lattice* (Sarkar, 2008) (see Figure 3.11). Function `grey.colors()` creates a vector of gamma-corrected gray colors. A similar function is `rainbow()`, which creates a vector of contiguous colors.

```
> library(lattice)

> levelplot(Petal.Width~Sepal.Length*Sepal.Width, iris, cuts=9,

+            col.regions=grey.colors(10)[10:1])
```

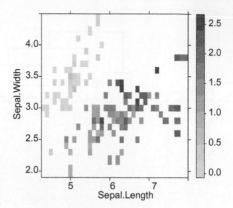

Figure 3.11 Level plot.

Contour plots can be plotted with `contour()` and `filled.contour()` in package *graphics*, and with `contourplot()` in package *lattice* (see Figure 3.12).

```
> filled.contour(volcano, color=terrain.colors, asp=1

+                  plot.axes=contour(volcano, add=T))
```

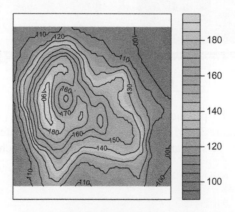

Figure 3.12 Contour.

Another way to illustrate a numeric matrix is a 3D surface plot shown as below, which is generated with function `persp()` (see Figure 3.13).

```
> persp(volcano, theta=25, phi=30, expand=0.5,
  col="lightblue")
```

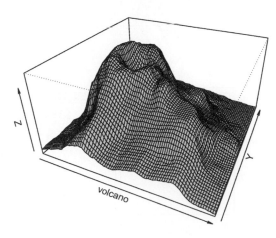

Figure 3.13 3D surface.

Parallel coordinates provide nice visualization of multiple dimensional data. A parallel coordinates plot can be produced with parcoord() in package *MASS*, and with parallelplot() in package *lattice* (see Figures 3.14 and 3.15).

```
> library(MASS)

> parcoord(iris[1:4], col=iris$Species)
```

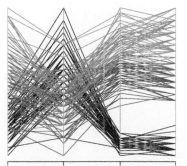

Sepal.Length Sepal.Width Petal.Length Petal.Width

Figure 3.14 Parallel coordinates.

```
> library(lattice)

> parallelplot(~iris[1:4] / Species, data=iris)
```

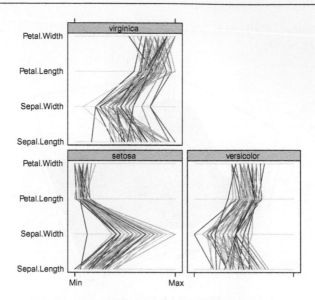

Figure 3.15 Parallel coordinates with package *lattice*.

Package *ggplot2* (Wickham, 2009) supports complex graphics, which are very useful for exploring data. A simple example is given below (see Figure 3.16). More examples on that package can be found at `http://had.co.nz/ggplot2/`.

```
> library(ggplot2)

> qplot(Sepal.Length, Sepal.Width, data=iris, facets=Species ~.)
```

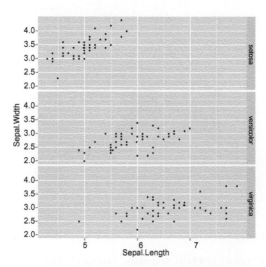

Figure 3.16 Scatter plot with package *ggplot2*.

3.5   Save Charts into Files

If there are many graphs produced in data exploration, a good practice is to save them into files. R provides a variety of functions for that purpose. Below are examples of saving charts into PDF and PS files respectively with pdf () and postscript (). Picture files of BMP, JPEG, PNG, and TIFF formats can be generated respectively with bmp (), jpeg (), png (), and tiff (). Note that the files (or graphics devices) need to be closed with graphics.off () or dev.off () after plotting.

```
> # save as a PDF file

> pdf("myPlot.pdf")

> x <- 1:50

> plot(x, log(x))

> graphics.off()

> #

> # save as a postscript file

> postscript("myPlot2.ps")

> x <- -20:20

> plot(x, x^2)

> graphics.off()
```

4 Decision Trees and Random Forest

This chapter shows how to build predictive models with packages *party*, *rpart* and *randomForest*. It starts with building decision trees with package *party* and using the built tree for classification, followed by another way to build decision trees with package *rpart*. After that, it presents an example on training a random forest model with package *randomForest*.

4.1 Decision Trees with Package *party*

This section shows how to build a decision tree for the `iris` data with function `ctree()` in package *party* (Hothorn et al., 2010). Details of the data can be found in Section 1.3.1. `Sepal.Length`, `Sepal.Width`, `Petal.Length`, and `Petal.Width` are used to predict the `Species` of flowers. In the package, function `ctree()` builds a decision tree, and `predict()` makes prediction for new data.

Before modeling, the `iris` data is split below into two subsets: training (70%) and test (30%). The random seed is set to a fixed value below to make the results reproducible.

```
> str(iris)

'data.frame':  150 obs. of 5 variables:

$ Sepal.Length: num 5.1 4.9 4.7 4.6 5 5.4 4.6 5 4.4 4.9 …

$ Sepal.Width: num 3.5 3 3.2 3.1 3.6 3.9 3.4 3.4 2.9 3.1 …

$ Petal.Length: num 1.4 1.4 1.3 1.5 1.4 1.7 1.4 1.5 1.4 1.5 …

$ Petal.Width: num 0.2 0.2 0.2 0.2 0.2 0.4 0.3 0.2 0.2 0.1 …

$ Species: Factor w/ 3 levels "setosa","versicolor",..: 1 1 1 1 1
  1 1 1 1 …

> set.seed(1234)

> ind <- sample(2, nrow(iris), replace=TRUE, prob=c(0.7, 0.3))
```

R and Data Mining. DOI: http://dx.doi.org/10.1016/B978-0-12-396963-7.00004-0

ℛⅅℳ

```
> trainData <- iris[ind==1,]

> testData <- iris[ind==2,]
```

We then load package *party*, build a decision tree, and check the prediction result. Function `ctree()` provides some parameters, such as `MinSplit`, `MinBusket`, `MaxSurrogate`, and `MaxDepth`, to control the training of decision trees. Below we use default settings to build a decision tree. Examples of setting the above parameters are available in Chapter 13. In the code below, `myFormula` specifies that `Species` is the target variable and all other variables are independent variables.

```
> library(party)

> myFormula <- Species ~ Sepal.Length + Sepal.Width +
  Petal.Length + Petal.Width

> iris_ctree <- ctree(myFormula, data=trainData)

> # check the prediction

> table(predict(iris_ctree), trainData$Species)
```

	setosa	versicolor	virginica
setosa	40	0	0
versicolor	0	37	3
virginica	0	1	31

After that, we can have a look at the built tree by printing the rules and plotting the tree.

```
> print(iris_ctree)

   Conditional inference tree with 4 terminal nodes

Response: Species

Inputs: Sepal.Length, Sepal.Width, Petal.Length, Petal.Width

Number of observations: 112

1) Petal.Length <= 1.9; criterion = 1, statistic = 104.643

  2)* weights = 40
```

```
1) Petal.Length > 1.9

  3) Petal.Width <= 1.7; criterion = 1, statistic = 48.939

    4) Petal.Length <= 4.4; criterion = 0.974, statistic = 7.397

      5)* weights = 21

    4) Petal.Length > 4.4

      6)* weights = 19

  3) Petal.Width > 1.7

      7)* weights = 32
```
> *plot(iris_ctree)*

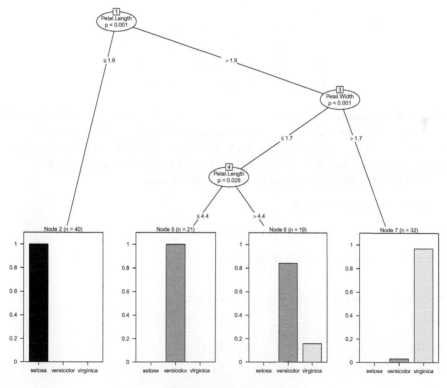

Figure 4.1 Decision tree.

> *plot(iris_ctree, type="simple")*

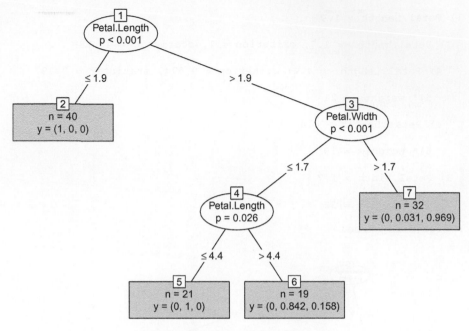

Figure 4.2 Decision tree (simple style).

In the above Figure 4.1, the barplot for each leaf node shows the probabilities of an instance falling into the three species. In Figure 4.2, they are shown as "y" in leaf nodes. For example, node 2 is labeled with "n = 40, y = (1, 0, 0)," which means that it contains 40 training instances and all of them belong to the first class "setosa."

After that, the built tree needs to be tested with test data.

```
> # predict on test data

> testPred <- predict(iris_ctree, newdata = testData)

> table(testPred, testData$Species)
```

testPred	setosa	versicolor	virginica
setosa	10	0	0
versicolor	0	12	2
virginica	0	0	14

The current version of `ctree()` (i.e. version 0.9-9995) does not handle missing values well, in that an instance with a missing value may sometimes go to the left sub-tree and sometimes to the right. This might be caused by surrogate rules.

Another issue is that, when a variable exists in training data and is fed into `ctree()` but does not appear in the built decision tree, the test data must also have that variable to make prediction. Otherwise, a call to `predict()` would fail. Moreover, if the value levels of a categorical variable in test data are different from that in train data, it would also fail to make prediction on the test data. One way to get around the above issue is, after building a decision tree, to call `ctree()` to build a new decision tree with data containing only those variables existing in the first tree, and to explicitly set the levels of categorical variables in test data to the levels of the corresponding variables in training data. An example on that can be found in Section 13.7.

4.2   Decision Trees with Package *rpart*

Package *rpart* (Therneau et al., 2010) is used in this section to build a decision tree on the `bodyfat` data (see Section 1.3.2 for details of the data). Function `rpart()` is used to build a decision tree, and the tree with the minimum prediction error is selected. After that, it is applied to new data to make prediction with function `predict()`.

At first, we load the `bodyfat` data and have a look at it.

```
> data("bodyfat", package = "mboost")

> dim(bodyfat)

[1] 71 10

> attributes(bodyfat)

$names

[1]    "age"          "DEXfat"    "waistcirc" "hipcirc"  "elbowbreadth"

[6]    "kneebreadth" "anthro3a"  "anthro3b"  "anthro3c" "anthro4"


$row.names

 [1]      "47" "48" "49" "50" "51" "52" "53" "54" "55" "56" "57" "58"

[13]      "59" "60" "61" "62" "63" "64" "65" "66" "67" "68" "69" "70"

[25]      "71" "72" "73" "74" "75" "76" "77" "78" "79" "80" "81" "82"

[37]      "83" "84" "85" "86" "87" "88" "89" "90" "91" "92" "93" "94"

[49]      "95" "96" "97" "98" "99" "100""101""102""103""104""105""106"

[61]      "107""108""109""110""111""112""113""114""115""116""117"
```

```
$class

[1] "data.frame"

> bodyfat[1:5,]
```

	age	DEXfat	waistcirc	hipcirc	elbowbreadth	kneebreadth	anthro3a
47	57	41.68	100.0	112.0	7.1	9.4	4.42
48	65	43.29	99.5	116.5	6.5	8.9	4.63
49	59	35.41	96.0	108.5	6.2	8.9	4.12
50	58	22.79	72.0	96.5	6.1	9.2	4.03
51	60	36.42	89.5	100.5	7.1	10.0	4.24

anthro3b	anthro3c	anthro4
4.95	4.50	6.13
5.01	4.48	6.37
4.74	4.60	5.82
4.48	3.91	5.66
4.68	4.15	5.91

Next, the data is split into training and test subsets, and a decision tree is built on the training data.

```
> set.seed(1234)

> ind <- sample(2, nrow(bodyfat), replace = TRUE, prob = c(0.7,
  0.3))

> bodyfat.train <- bodyfat[ind==1,]

> bodyfat.test <- bodyfat[ind==2,]

> # train a decision tree

> library(rpart)

> myFormula <- DEXfat ~age + waistcirc + hipcirc + elbowbreadth +
  kneebreadth

> bodyfat_rpart <- rpart(myFormula, data = bodyfat.train,

+                            control = rpart.control(minsplit = 10))

> attributes(bodyfat_rpart)
```

```
$names
[1]     "frame"  "where" "call"    "terms"     "cptable" "splits"
[7]     "method" "parms" "control" "functions" "y"       "ordered"
$class
[1] "rpart"
```

> *print(bodyfat_rpart$cptable)*

	CP	nsplit	rel error	xerror	xstd
1	0.67272638	0	1.00000000	1.0194546	0.18724382
2	0.09390665	1	0.32727362	0.4415438	0.10853044
3	0.06037503	2	0.23336696	0.4271241	0.09362895
4	0.03420446	3	0.17299193	0.3842206	0.09030539
5	0.01708278	4	0.13878747	0.3038187	0.07295556
6	0.01695763	5	0.12170469	0.2739808	0.06599642
7	0.01007079	6	0.10474706	0.2693702	0.06613618
8	0.01000000	7	0.09467627	0.2695358	0.06620732

> *print(bodyfat_rpart)*

```
n = 56

node), split, n, deviance, yval
    * denotes terminal node

1) root 56 7265.0290000 30.94589
  2) waistcirc< 88.4 31 960.5381000 22.55645
    4) hipcirc< 96.25 14 222.2648000 18.41143
      8) age< 60.5 9 66.8809600 16.19222 *
      9) age>=60.5 5 31.2769200 22.40600 *
    5) hipcirc>=96.25 17 299.6470000 25.97000
      10) waistcirc< 77.75 6 30.7345500 22.32500 *
      11) waistcirc>=77.75 11 145.7148000 27.95818
        22) hipcirc< 99.5 3 0.2568667 23.74667 *
        23) hipcirc>=99.5 8 72.2933500 29.53750 *
```

```
3) waistcirc>=88.4 25 1417.1140000 41.34880

    6) waistcirc< 104.75 18 330.5792000 38.09111

        12) hipcirc< 109.9 9 68.9996200 34.37556 *

        13) hipcirc>=109.9 9 13.0832000 41.80667 *

    7) waistcirc>=104.75 7 404.3004000 49.72571 *
```

The build tree can be plotted with the code below (see Figure 4.3).

```
> plot(bodyfat_rpart)

> text(bodyfat_rpart, use.n=T)
```

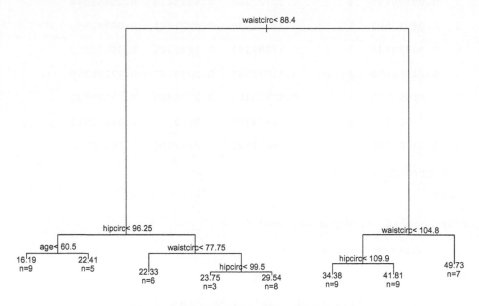

Figure 4.3 Decision tree with package *rpart*.

Then we select the tree with the minimum prediction error (see Figure 4.4).

```
> opt <- which.min(bodyfat_rpart$cptable[,"xerror"])

> cp <- bodyfat_rpart$cptable[opt, "CP"]

> bodyfat_prune <- prune(bodyfat_rpart, cp = cp)

> print(bodyfat_prune)

n = 56

node), split, n, deviance, yval

    * denotes terminal node
```

```
1) root 56 7265.02900 30.94589

  2) waistcirc< 88.4 31 960.53810 22.55645

    4) hipcirc< 96.25 14 222.26480 18.41143

      8) age< 60.5 9 66.88096 16.19222 *

      9) age>=60.5 5 31.27692 22.40600 *

    5) hipcirc>=96.25 17 299.64700 25.97000

      10) waistcirc< 77.75 6 30.73455 22.32500 *

      11) waistcirc>=77.75 11 145.71480 27.95818 *

  3) waistcirc>=88.4 25 1417.11400 41.34880

    6) waistcirc< 104.75 18 330.57920 38.09111

      12) hipcirc< 109.9 9 68.99962 34.37556 *

      13) hipcirc>=109.9 9 13.08320 41.80667 *

    7) waistcirc>=104.75 7 404.30040 49.72571 *

> plot(bodyfat_prune)

> text(bodyfat_prune, use.n=T)
```

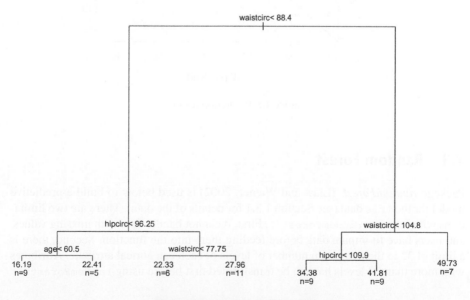

Figure 4.4 Selected decision tree.

After that, the selected tree is used to make prediction and the predicted values are compared with actual labels. In the code below, function `abline()` draws a diagonal line. The predictions of a good model are expected to be equal or very close to their actual values, that is, most points should be on or close to the diagonal line (see Figure 4.5).

```
> DEXfat_pred <- predict(bodyfat_prune, newdata=bodyfat.test)

> xlim <- range(bodyfat$DEXfat)

> plot(DEXfat_pred ~ DEXfat, data=bodyfat.test, xlab="Observed",

+       ylab="Predicted", ylim=xlim, xlim=xlim)

> abline(a=0, b=1)
```

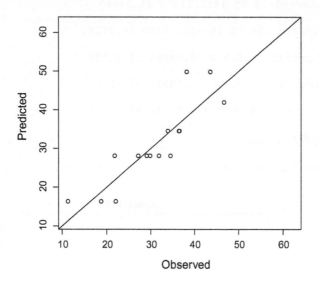

Figure 4.5 Prediction result.

4.3   Random Forest

Package *randomForest* (Liaw and Wiener, 2002) is used below to build a predictive model for the `iris` data (see Section 1.3.1 for details of the data). There are two limitations with function `randomForest()`. First, it cannot handle data with missing values, and users have to impute data before feeding them into the function. Second, there is a limit of 32 to the maximum number of levels of each categorical attribute. Attributes with more than 32 levels have to be transformed first before using `randomForest()`.

An alternative way to build a random forest is to use function cforest() from package *party*, which is not limited to the above maximum levels. However, generally speaking, categorical variables with more levels will make it require more memory and take longer time to build a random forest.

Again, the iris data is first split into two subsets: training (70%) and test (30%).

```
> ind <- sample(2, nrow(iris), replace=TRUE, prob=c(0.7, 0.3))

> trainData <- iris[ind==1,]

> testData <- iris[ind==2,]
```

Then we load package *randomForest* and train a random forest. In the code below, the formula is set to "Species ~.", which means to predict Species with all other variables in the data.

```
> library(randomForest)

> rf <- randomForest(Species ~ ., data=trainData, ntree=100,
  proximity=TRUE)

> table(predict(rf), trainData$Species)
```

	setosa	versicolor	virginica
setosa	36	0	0
versicolor	0	31	2
virginica	0	1	34

```
> print(rf)

Call:
    randomForest(formula = Species ~ ., data = trainData,
       ntree = 100, proximity = TRUE)

          Type of random forest: classification

               Number of trees: 100

No. of variables tried at each split: 2

          OOB estimate of error rate: 2.88%
```

```
Confusion matrix:
           setosa    versicolor    virginica    class.error
setosa     36        0             0            0.00000000
versicolor 0         31            1            0.03125000
virginica  0         2             34           0.05555556
```

```
> attributes(rf)
$names
[1]    "call"          "type"           "predicted"        "err.rate"
[5]    "confusion"     "votes"          "oob.times"        "classes"
[9]    "importance"    "importanceSD"   "localImportance"  "proximity"
[13]   "ntree"         "mtry"           "forest"           "y"
[17]   "test"          "inbag"          "terms"

$class
[1] "randomForest.formula" "randomForest"
```

After that, we plot the error rates with various number of trees (see Figure 4.6).

```
> plot(rf)
```

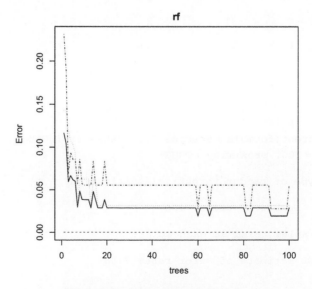

Figure 4.6 Error rate of random forest.

The importance of variables can be obtained with functions `importance()` and `varImpPlot()` (see Figure 4.7).

```
> importance(rf)
```

```
                MeanDecreaseGini
Sepal.Length    6.913882
Sepal.Width     1.282567
Petal.Length    26.267151
Petal.Width     34.163836
```

```
> varImpPlot(rf)
```

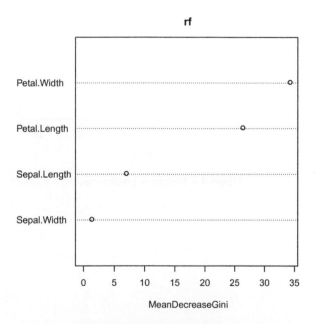

Figure 4.7 Variable importance.

Finally, the built random forest is tested on test data, and the result is checked with functions `table()` and `margin()` (see Figure 4.8). The margin of a data point is the proportion of votes for the correct class minus maximum proportion of votes for other classes. Generally speaking, positive margin means correct classification.

```
> irisPred <- predict(rf, newdata=testData)
```

```
> table(irisPred, testData$Species)
```

```
irisPred     setosa   versicolor   virginica

setosa       14       0            0

versicolor   0        17           3

virginica    0        1            11
```

```
> plot(margin(rf, testData$Species))
```

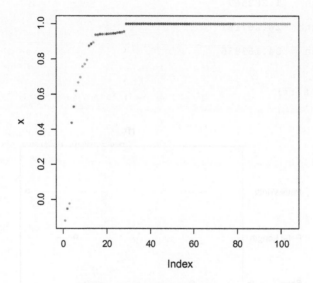

Figure 4.8 Margin of predictions.

5 Regression

Regression is to build a function of *independent variables* (also known as *predictors*) to predict a *dependent variable* (also called *response*). For example, banks assess the risk of home-loan applicants based on their age, income, expenses, occupation, number of dependents, total credit limit, etc.

This chapter introduces basic concepts and presents examples of various regression techniques. At first, it shows an example on building a linear regression model to predict CPI data. After that, it introduces logistic regression. The generalized linear model (GLM) is then presented, followed by a brief introduction of non-linear regression.

A collection of some helpful R functions for regression analysis is available as a reference card on *R Functions for Regression Analysis*.[1]

5.1 Linear Regression

Linear regression is to predict response with a linear function of predictors as follows:

$$y = c_0 + c_1 x_1 + c_2 x_2 + \cdots + c_k x_k,$$

where $x_1, x_2, \cdots, x_k$ are predictors and y is the response to predict.

Linear regression is demonstrated below with function `lm()` on the Australian CPI (Consumer Price Index) data, which are quarterly CPIs from 2008 to 2010.[2]

At first, the data is created and plotted. In the code below, an x-axis is added manually with function `axis()`, where `las = 3` makes text vertical (see Figure 5.1).

```
> year <- rep(2008:2010, each = 4)

> quarter <- rep(1:4, 3)

> cpi <- c(162.2, 164.6, 166.5, 166.0,

+          166.2, 167.0, 168.6, 169.5,

+          171.0, 172.1, 173.3, 174.0)

> plot(cpi, xaxt="n", ylab="CPI", xlab="")
```

---

[1] http://cran.r-project.org/doc/contrib/Ricci-refcard-regression.pdf
[2] From Australian Bureau of Statistics http://www.abs.gov.au

R and Data Mining. DOI: http://dx.doi.org/10.1016/B978-0-12-396963-7.00005-2

```
> # draw x-axis

> axis(1, labels=paste(year,quarter,sep="Q"), at=1:12, las=3)
```

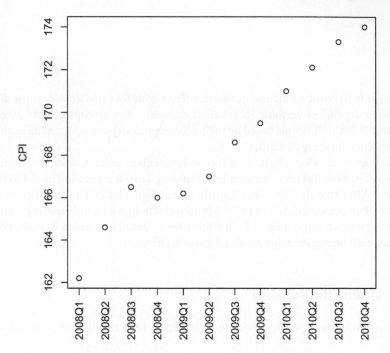

Figure 5.1 Australian CPIs in year 2008 to 2010.

We then check the correlation between CPI and the other variables, year and quarter.

```
> cor(year,cpi)

[1] 0.9096316

> cor(quarter,cpi)

[1] 0.3738028
```

Then a linear regression model is built with function lm() on the above data, using year and quarter as predictors and CPI as response.

```
> fit <- lm(cpi˜ year + quarter)

> fit
```

```
Call:

lm(formula = cpi~ year + quarter)

Coefficients:

 (Intercept)    year      quarter
  -7644.488    3.888     1.167
```

With the above linear model, CPI is calculated as

$$\mathrm{cpi} = c_0 + c_1 * \mathrm{year} + c_2 * \mathrm{quarter},$$

where c_0, c_1, and c_2 are coefficients from model `fit`. Therefore, the CPIs in 2011 can be calculated as follows. An easier way for this is using function `predict()`, which will be demonstrated at the end of this subsection.

```
> (cpi2011 <- fit$coefficients[[1]] + fit$coefficients[[2]]*2011 +

+                fit$coefficients[[3]]*(1:4))

[1] 174.4417 175.6083 176.7750 177.9417
```

More details of the model can be obtained with the code below.

```
> attributes(fit)

$names
 [1]      "coefficients"    "residuals"    "effects"    "rank"
 [5]      "fitted.values"   "assign"       "qr"         "df.residual"
 [9]      "xlevels"         "call"         "terms"      "model"

$class
 [1] "lm"

> fit$coefficients

 (Intercept)      year         quarter
  -7644.487500   3.887500    1.166667
```

   The differences between observed values and fitted values can be obtained with
function residuals().

```
> # differences between observed values and fitted values
> residuals(fit)
```

```
            1            2          3           4           5           6
-0.57916667  0.65416667 1.38750000 -0.27916667 -0.46666667 -0.83333333
            7            8          9          10          11          12
-0.40000000 -0.66666667 0.44583333  0.37916667  0.41250000 -0.05416667
```

```
> summary(fit)

Call:

lm(formula = cpi~ year + quarter)

Residuals:

Min       1Q       Median     3Q       Max
-0.8333   -0.4948  -0.1667    0.4208   1.3875
```

Coefficients:

	Estimate	Std. Error	t value	Pr(>|t|)	
(Intercept)	-7644.4875	518.6543	-14.739	1.31e-07	***
year	3.8875	0.2582	15.058	1.09e-07	***
quarter	1.1667	0.1885	6.188	0.000161	***

```
---

Signif. codes:  0 ś*** š   0.001 ś**š  0.01 ś*š  0.05 ś.š  0.1śš1
```

```
Residual standard error: 0.7302 on 9 degrees of freedom

Multiple R-squared: 0.9672,    Adjusted R-squared: 0.9599

F-statistic: 132.5 on 2 and 9 DF, p-value: 2.108e-07
```

We then plot the fitted model with the following code (see Figure 5.2).

```
> plot(fit)
```

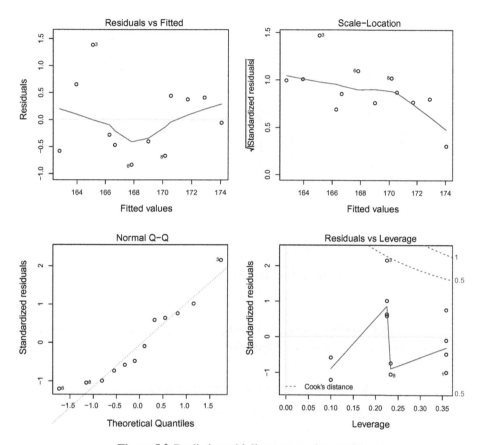

Figure 5.2 Prediction with linear regression model.

We can also plot the model in a 3D plot as below, where function `scatterplot3d()` creates a 3D scatter plot and `plane3d()` draws the fitted plane. Parameter `lab` specifies the number of tickmarks on the x- and y-axes (see Figure 5.3).

```
> library(scatterplot3d)

> s3d <- scatterplot3d(year, quarter, cpi, highlight.3d=T,
  type="h", lab=c(2,3))
```

```
> s3d$plane3d(fit)
```

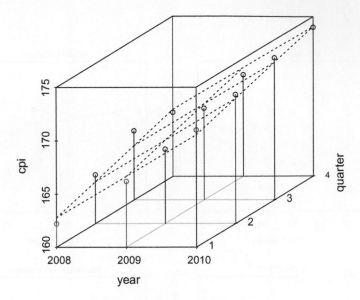

Figure 5.3 A 3D plot of the fitted model.

With the model, the CPIs in year 2011 can be predicted as follows, and the predicted values are shown as red triangles in Figure 5.4.

```
> data2011 <- data.frame(year=2011, quarter=1:4)

> cpi2011 <- predict(fit, newdata=data2011)

> style <- c(rep(1,12), rep(2,4))

> plot(c(cpi, cpi2011), xaxt="n", ylab="CPI", xlab="",
  pch = style, col = style)

> axis(1, at=1:16, las=3,

+       labels=c(paste(year,quarter,sep="Q"), "2011Q1", "2011Q2",
  "2011Q3", "2011Q4"))
```

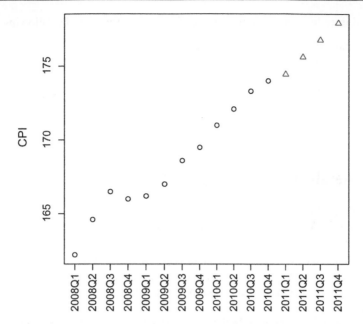

Figure 5.4 Prediction of CPIs in 2011 with linear regression model.

5.2 Logistic Regression

Logistic regression is used to predict the probability of occurrence of an event by fitting data to a logistic curve. A logistic regression model is built as the following equation:

$$logit(y) = c_0 + c_1 x_1 + c_2 x_2 + \cdots + c_k x_k,$$

where $x_1, x_2, \cdots, x_k$ are predictors, y is a response to predict, and $logit(y) = ln\left(\frac{y}{1-y}\right)$. The above equation can also be written as

$$y = \frac{1}{1 + e^{-(c_0 + c_1 x_1 + c_2 x_2 + \cdots + c_k x_k)}}.$$

Logistic regression can be built with function `glm()` by setting `family` to `binomial(link="logit")`.

Detailed introductions on logistic regression can be found at the following links.

• R Data Analysis Examples—Logit Regression
  http://www.ats.ucla.edu/stat/r/dae/logit.htm

• Logistic Regression (with R)
  http://nlp.stanford.edu/~manning/courses/ling289/logistic.pdf

5.3 Generalized Linear Regression

The generalized linear model (GLM) generalizes linear regression by allowing the linear model to be related to the response variable via a link function and allowing the magnitude of the variance of each measurement to be a function of its predicted value. It unifies various other statistical models, including linear regression, logistic regression, and Poisson regression. Function glm() is used to fit generalized linear models, specified by giving a symbolic description of the linear predictor and a description of the error distribution.

A generalized linear model is built below with glm() on the bodyfat data (see 1.3.2 for details of the data).

```
> data("bodyfat", package = "mboost")

> myFormula <- DEXfat ~age + waistcirc + hipcirc + elbowbreadth +
  kneebreadth

> bodyfat.glm <- glm(myFormula, family = gaussian("log"),
  data = bodyfat)

> summary(bodyfat.glm)
Call:

glm(formula = myFormula, family = gaussian("log"), data = bodyfat)

Deviance Residuals:

Min        1Q       Median   3Q       Max

-11.5688   -3.0065   0.1266   2.8310   10.0966
```

Coefficients:

	Estimate	Std. Error	t value	Pr(>|t|)	
(Intercept)	0.734293	0.308949	2.377	0.02042	*
age	0.002129	0.001446	1.473	0.14560	
waistcirc	0.010489	0.002479	4.231	7.44e-05	***
hipcirc	0.009702	0.003231	3.003	0.00379	**
elbowbreadth	0.002355	0.045686	0.052	0.95905	
kneebreadth	0.063188	0.028193	2.241	0.02843	*

—

Signif. codes: 0 ś***š   0.001 ś**š 0.01 ś*š 0.05 ś.š 0.1 ś š 1

(Dispersion parameter for gaussian family taken to be 20.31433)

Null deviance: 8536.0   on 70 degrees of freedom

Residual deviance: 1320.4   on 65 degrees of freedom

AIC: 423.02

Number of Fisher Scoring iterations: 5

```
> pred <- predict(bodyfat.glm, type="response")
```

In the code above, `type` indicates the type of prediction required. The default is on the scale of the linear predictors, and the alternative `"response"` is on the scale of the response variable. We then plot the predicted result with the code below (see Figure 5.5).

```
> plot(bodyfat$DEXfat, pred, xlab="Observed Values",
  ylab="Predicted Values")

> abline(a=0, b=1)
```

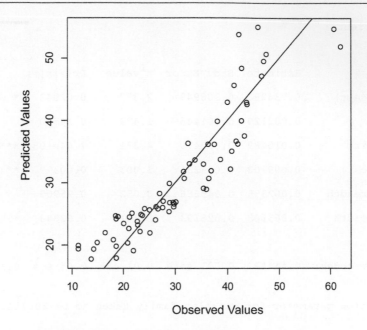

Figure 5.5 Prediction with generalized linear regression model.

In the above code, if `family = gaussian("identity")` is used, the built model would be similar to linear regression. One can also make it a logistic regression by setting `family` to `binomial("logit")`.

5.4 Non-Linear Regression

While linear regression is to find the line that comes closest to data, non-linear regression is to fit a curve through data. Function `nls()` provides non-linear regression. Examples of `nls()` can be found by running `"?nls"` under R.

6 Clustering

This chapter presents examples of various clustering techniques in *R*, including *k*-means clustering, *k*-medoids clustering, hierarchical clustering, and density-based clustering. The first two sections demonstrate how to use the *k*-means and *k*-medoids algorithms to cluster the *iris* data. The third section shows an example on hierarchical clustering on the same data. The last section describes the idea of density-based clustering and the DBSCAN algorithm, and shows how to cluster with DBSCAN and then label new data with the clustering model. For readers who are not familiar with clustering, introductions of various clustering techniques can be found in Zhao et al. (2009a) and Jain et al. (1999).

6.1 The k-Means Clustering

This section shows k-means clustering of `iris` data (see Section 1.3.1 for details of the data). At first, we remove species from the data to cluster. After that, we apply function `kmeans()` to iris2, and store the clustering result in `kmeans.result`. The cluster number is set to 3 in the code below.

```
> iris2 <- iris

> iris2$Species <- NULL

> (kmeans.result <- kmeans(iris2, 3))

K-means clustering with 3 clusters of sizes 38, 50, 62

Cluster means:

   Sepal.Length  Sepal.Width  Petal.Length  Petal.Width

1  6.850000      3.073684     5.742105      2.071053

2  5.006000      3.428000     1.462000      0.246000

3  5.901613      2.748387     4.393548      1.433871


Clustering vector:

 [1] 2 2 2 2 2 2 2 2 2 2 2 2 2 2 2 2 2 2 2 2 2 2 2 2 2 2 2 2 2 2 2 2 2 2 2 2 2

[38] 2 2 2 2 2 2 2 2 2 2 2 2 2 3 3 1 3 3 3 3 3 3 3 3 3 3 3 3 3 3 3 3 3 3 3 3 3
```

R and Data Mining. DOI: http://dx.doi.org/10.1016/B978-0-12-396963-7.00006-4

```
[75] 3 3 3 1 3 3 3 3 3 3 3 3 3 3 3 3 3 3 3 3 3 3 3 3 3 3 3 3 3 1 3 1 1 1 1 3 1 1 1 1

[112] 1 1 3 3 1 1 1 1 3 1 3 1 3 1 3 1 1 3 3 1 1 1 1 3 1 1 1 1 3 1 1 1 3 1 1 1 3 1

[149] 1 3
```

```
Within cluster sum of squares by cluster:

[1] 23.87947 15.15100 39.82097

  (between_SS / total_SS = 88.4%)

Available components:

 [1]  "cluster"      "centers"     "totss"      "withinss"    "tot.withinss"

 [6]  "betweenss"    "size"
```

The clustering result is then compared with the class label (Species) to check whether similar objects are grouped together.

```
> table(iris$Species, kmeans.result$cluster)
```

	1	2	3
setosa	0	50	0
versicolor	2	0	48
virginica	36	0	14

The above result shows that cluster "setosa" can be easily separated from the other clusters, and that clusters "versicolor" and "virginica" are to a small degree overlapped with each other.

Next, the clusters and their centers are plotted (see Figure 6.1). Note that there are four dimensions in the data and that only the first two dimensions are used to draw the plot below. Some black points close to the green center (asterisk) are actually closer to the black center in the four-dimensional space. We also need to be aware that the results of k-means clustering may vary from run to run, due to random selection of initial cluster centers.

```
> plot(iris2[c("Sepal.Length", "Sepal.Width")],
  col = kmeans.result$cluster

> # plot cluster centers

> points(kmeans.result$centers[,c("Sepal.Length",
  "Sepal.Width")], col=1:3, pch=8, cex=2)
```

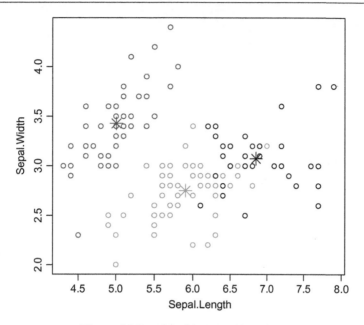

Figure 6.1 Results of k-means clustering.

More examples of k-means clustering can be found in Section 7.3 and Section 10.8.1.

6.2   The k-Medoids Clustering

This section shows k-medoids clustering with functions `pam()` and `pamk()`. The k-medoids clustering is very similar to k-means, and the major difference between them is that: while a cluster is represented with its center in the k-means algorithm, it is represented with the object closest to the center of the cluster in the k-medoids clustering. The k-medoids clustering is more robust than k-means in presence of outliers. PAM (Partitioning Around Medoids) is a classic algorithm for k-medoids clustering. While the PAM algorithm is inefficient for clustering large data, the CLARA algorithm is an enhanced technique of PAM by drawing multiple samples of data, applying PAM on each sample and then returning the best clustering. It performs better than PAM on larger data. Functions `pam()` and `clara()` in package *cluster* (Maechler et al., 2012) are respectively implementations of PAM and CLARA in R. For both algorithms, a user has to specify k, the number of clusters to find. As an enhanced version of `pam()`, function `pamk()` in package *fpc* (Hennig, 2010) does not require a user to choose k. Instead, it calls the function `pam()` or `clara()` to perform a partitioning around medoids clustering with the number of clusters estimated by optimum average silhouette width.

    With the code below, we demonstrate how to find clusters with `pam()` and `pamk()`.

```
> library(fpc)
```

```
> pamk.result <- pamk(iris2)

> # number of clusters

> pamk.result$nc

[1] 2

> # check clustering against actual species

> table(pamk.result$pamobject$clustering, iris$Species)

      setosa   versicolor   virginica

  1   50        1            0

  2   0         49           50

> layout(matrix(c(1,2),1,2)) # 2 graphs per page

> plot(pamk.result$pamobject)

> layout(matrix(1)) # change back to one graph per page
```

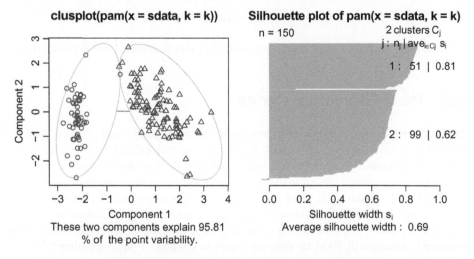

Figure 6.2 Clustering with the k-medoids algorithm—I.

In the above example, pamk() produces two clusters: one is "setosa", and the other is a mixture of "versicolor" and "virginica". In Figure 6.2, the left-side chart is a two-dimensional "clusplot" (clustering plot) of the two clusters and the lines show the distance between clusters. The right one shows their silhouettes. In the silhouette, a large s_i (almost 1) suggests that the corresponding observations are very well clustered, a small s_i (around 0) means that the observation lies between two clusters, and observations with a negative s_i are probably placed in the wrong cluster. Since the average S_i are respectively 0.81 and 0.62 in the above silhouette, the identified two clusters are well clustered.

Next, we try pam() with $k = 3$.

```
> pam.result <- pam(iris2, 3)

> table(pam.result$clustering, iris$Species)

    setosa   versicolor   virginica

1   50       0            0

2   0        48           14

3   0        2            36

> layout(matrix(c(1,2),1,2)) # 2 graphs per page

> plot(pam.result)

> layout(matrix(1)) # change back to one graph per page
```

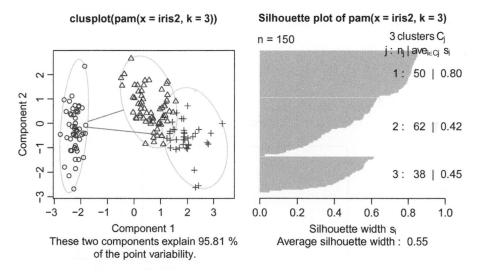

Figure 6.3 Clustering with the k-medoids algorithm—II.

With the above result produced with pam() (see Figure 6.3), there are three clusters:
(1) cluster 1 is species "setosa" and is well separated from the other two; (2) cluster
2 is mainly composed of "versicolor", plus some cases from "virginica"; and (3) the
majority of cluster 3 are "virginica", with two cases from "versicolor".

It is hard to say which one is better out of the above two clusterings produced
respectively with pamk() and pam(). It depends on the target problem and domain
knowledge and experience. In this example, the result of pam() seems better, because
it identifies three clusters, corresponding to three species. Therefore, the heuristic way
to identify the number of clusters in pamk() does not necessarily produce the best
result. Note that we cheated by setting $k = 3$ when using pam(), which is already
known to us as the number of species.

More examples of k-medoids clustering can be found in Section 10.8.2.

6.3   Hierarchical Clustering

This section demonstrates hierarchical clustering with hclust() on iris data (see Section 1.3.1 for details of the data).

We first draw a sample of 40 records from the iris data, so that the clustering plot will not be overcrowded. Same as before, variable Species is removed from the data. After that, we apply hierarchical clustering to the data.

```
> idx <- sample(1:dim(iris)[1], 40)

> irisSample <- iris[idx,]

> irisSample$Species <- NULL

> hc <- hclust(dist(irisSample), method="ave")

> plot(hc, hang = -1, labels=iris$Species[idx])

> # cut tree into 3 clusters

> rect.hclust(hc, k=3)

> groups <- cutree(hc, k=3)
```

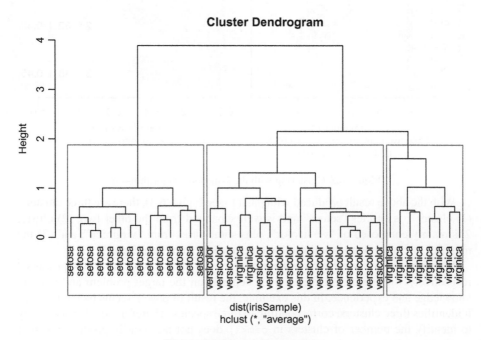

Figure 6.4 Cluster dendrogram.

Similar to the above clustering of *k*-means, Figure 6.4 also shows that cluster "setosa" can be easily separated from the other two clusters, and that clusters "versicolor" and "virginica" are to a small degree overlapped with each other.

More examples of hierarchical clustering can be found in Section 8.4 and Section 10.7.

6.4 Density-Based Clustering

The DBSCAN algorithm (Ester et al., 1996) from package *fpc* (Hennig, 2010) provides a density-based clustering for numeric data. The idea of density-based clustering is to group objects into one cluster if they are connected to one another by densely populated area. There are two key parameters in DBSCAN:

- `eps`: reachability distance, which defines the size of neighborhood; and

- `MinPts`: minimum number of points.

If the number of points in the neighborhood of point α is no less than `MinPts`, then α is a *dense point*. All the points in its neighborhood are *density-reachable* from α and are put into the same cluster as α.

The strengths of density-based clustering are that it can discover clusters with various shapes and sizes and is insensitive to noise. As a comparison, the *k*-means algorithm tends to find clusters with sphere shape and with similar sizes.

Below is an example of density-based clustering of the `iris` data.

```
> library(fpc)

> iris2 <- iris[-5] # remove class tags

> ds <- dbscan(iris2, eps=0.42, MinPts=5)

> # compare clusters with original class labels

> table(ds$cluster, iris$Species)
```

	setosa	versicolor	virginica
0	2	10	17
1	48	0	0
2	0	37	0
3	0	3	33

In the above table, "1" to "3" in the first column are three identified clusters, while
"0" stands for noises or outliers, i.e. objects that are not assigned to any clusters. The
noises are shown as black circles in Figure 6.5.

```
> plot(ds, iris2)
```

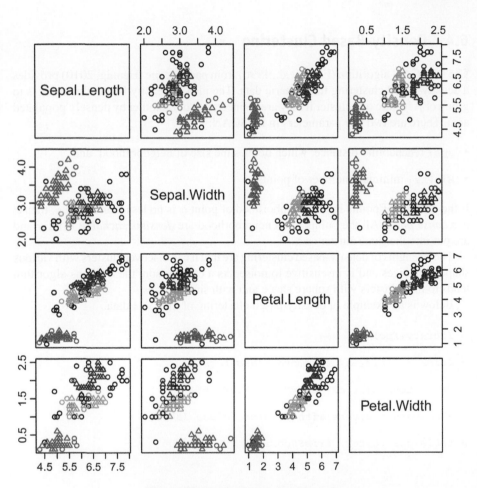

Figure 6.5 Density-based clustering—I.

The clusters are shown below in a scatter plot using the first and fourth columns of
the data (see Figure 6.6).

```
> plot(ds, iris2[c(1,4)])
```

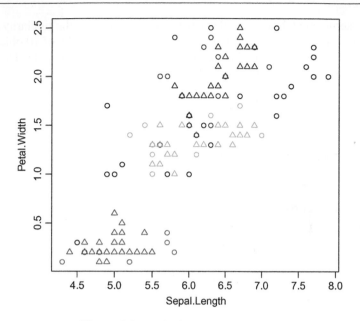

Figure 6.6 Density-based clustering—II.

Another way to show the clusters is using function `plotcluster()` in package *fpc*. Note that the data are projected to distinguish classes (see Figure 6.7.)

```
> plotcluster(iris2, ds$cluster)
```

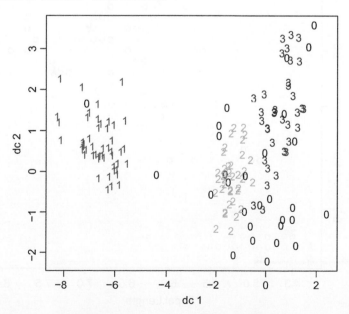

Figure 6.7 Density-based clustering—III.

The clustering model can be used to label new data, based on the similarity between new data and the clusters. The following example draws a sample of 10 objects from iris and adds small noises to them to make a new dataset for labeling. The random noises are generated with a uniform distribution using function runif().

```
> # create a new dataset for labeling

> set.seed(435)

> idx <- sample(1:nrow(iris), 10)

> newData <- iris[idx, -5]

> newData <- newData + matrix(runif(10*4. min=0, max=0.2),
  nrow=10, ncol=4)

> # label new data

> myPred <- predict(ds, iris2, newData)

> # plot result

> plot(iris2[c(1,4)], col=1+ds$cluster)

> points(newData[c(1,4)], pch="ast", col=1+myPred, cex=3)
```

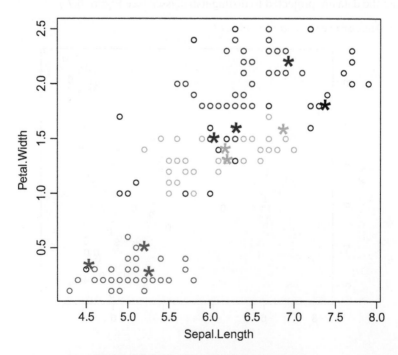

Figure 6.8 Prediction with clustering model.

```
> # check cluster labels

> table(myPred, iris$Species[idx])
```

mypred	setosa	versicolor	virginica
0	0	0	1
1	3	0	0
2	0	3	0
3	0	1	2

As we can see from the above result, out of the 10 new unlabeled data, $8 (= 3 + 3 + 2)$ are assigned with correct class labels. The new data are shown as asterisk("*") and the colors stand for cluster labels in Figure 6.8.

7 Outlier Detection

This chapter presents examples of outlier detection with R. At first, it demonstrates univariate outlier detection. After that, an example of outlier detection with LOF (Local Outlier Factor) is given, followed by examples on outlier detection by clustering. At last, it demonstrates outlier detection from time series data.

7.1 Univariate Outlier Detection

This section shows an example of univariate outlier detection and demonstrates how to apply it to multivariate data. In the example, univariate outlier detection is done with function `boxplot.stats()`, which returns the statistics for producing boxplots. In the result returned by the above function, one component is `out`, which gives a list of outliers. More specifically, it lists data points lying beyond the extremes of the whiskers. An argument of `coef` can be used to control how far the whiskers extend out from the box of a boxplot. More details on that can be obtained by running `?boxplot.stats` in R. Figure 7.1 shows a boxplot, where the four circles are outliers.

```
> set.seed(3147)

> x <- rnorm(100)

> summary(x)


  Min.    1st Qu.   Median   Mean    3rd Qu.   Max.

 -3.3150  -0.4837   0.1867   0.1098  0.7120    2.6860


> # outliers

> boxplot.stats(x)$out

[1] -3.315391 2.685922 -3.055717 2.571203

> boxplot(x)
```

R and Data Mining. DOI: http://dx.doi.org/10.1016/B978-0-12-396963-7.00007-6

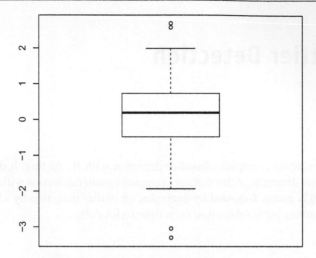

Figure 7.1 Univariate outlier detection with boxplot.

The above univariate outlier detection can be used to find outliers in multivariate data in a simple ensemble way. In the example below, we first generate a dataframe df, which has two columns, x and y. After that, outliers are detected separately from x and y. We then take outliers as those data which are outliers for both columns. In Figure 7.2, outliers are labeled with "+" in red..

```
> y <- rnorm(100)

> df <- data.frame(x, y)

> rm(x, y)

> head(df)

     x              y

1  -3.31539150    0.7619774

2  -0.04765067   -0.6404403

3   0.69720806    0.7645655

4   0.35979073    0.3131930

5   0.18644193    0.1709528

6   0.27493834   -0.8441813

> attach(df)

> # find the index of outliers from x

>(a <- which(x %in% boxplot.stats(x)$out))
```

```
[1] 1 33 64 74

> # find the index of outliers from y

>(b <- which(y %in% boxplot.stats(y)$out))

[1] 24 25 49 64 74

> detach(df)

> # outliers in both x and y

>(outlier.list1 <- intersect(a,b))

[1] 64 74

> plot(df)

> points(df[outlier.list1,], col="red", pch="+", cex=2.5)
```

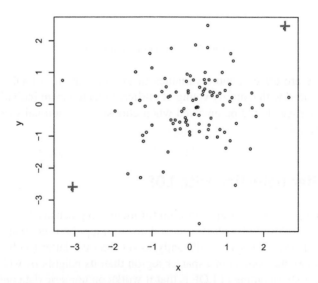

Figure 7.2 Outlier detection—I.

Similarly, we can also take outliers as those data which are outliers in either x or y. In Figure 7.3, outliers are labeled with "x" in blue.

```
> # outliers in either x or y

>(outlier.list2 <- union(a,b))

[1] 1 33 64 74 24 25 49

> plot(df)
```

```
> points(df[outlier.list2,], col="blue", pch="x", cex=2)
```

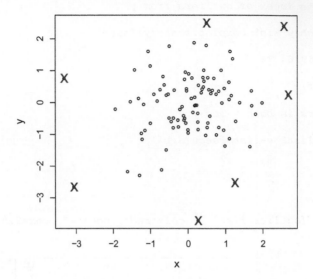

Figure 7.3 Outlier detection—II.

When there are three or more variables in an application, a final list of outliers might be produced with majority voting of outliers detected from individual variables. Domain knowledge should be involved when choosing the optimal way to ensemble in real-world applications.

7.2   Outlier Detection with LOF

LOF (Local Outlier Factor) is an algorithm for identifying density-based local outliers (Breunig et al., 2000). With LOF, the local density of a point is compared with that of its neighbors. If the former is significantly lower than the latter (with an LOF value greater than one), the point is in a sparser region than its neighbors, which suggests it be an outlier. A shortcoming of LOF is that it works on numeric data only.

Function lofactor() calculates local outlier factors using the LOF algorithm, and it is available in packages *DMwR* (Torgo, 2010) and *dprep*. An example of outlier detection with LOF is given below, where k is the number of neighbors used for calculating local outlier factors. Figure 7.4 shows a density plot of outlier scores.

```
> library(DMwR)

> # remove "Species", which is a categorical column

> iris2 <- iris[,1:4]

> outlier.scores <- lofactor(iris2, k=5)
```

```
> plot(density(outlier.scores))
```

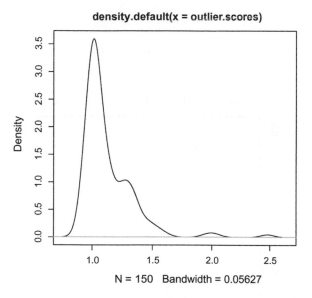

Figure 7.4 Density of outlier factors.

```
> # pick top 5 as outliers
> outliers <- order(outlier.scores, decreasing=T)[1:5]
> # who are outliers
> print(outliers)
[1] 42 107 23 110 63
> print(iris2[outliers,])
```

	Sepal.Length	Sepal.Width	Petal.Length	Petal.Width
42	4.5	2.3	1.3	0.3
107	4.9	2.5	4.5	1.7
23	4.6	3.6	1.0	0.2
110	7.2	3.6	6.1	2.5
63	6.0	2.2	4.0	1.0

Next, we show outliers with a biplot of the first two principal components (see Figure 7.5).

```
> n <- nrow(iris2)

> labels <- 1:n

> labels[-outliers] <- "."

> biplot(prcomp(iris2), cex=.8, xlabs=labels)
```

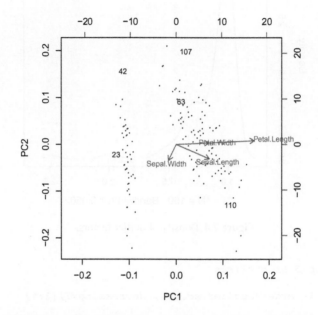

Figure 7.5 Outliers in a biplot of first two principal components.

In the above code, `prcomp()` performs a principal component analysis, and `biplot()` plots the data with its first two principal components. In Figure 7.5, the x- and y-axes are respectively the first and second principal components, the arrows show the original columns (variables), and the five outliers are labeled with their row numbers.

We can also show outliers with a pairs plot as below, where outliers are labeled with "+" in red (see Figure 7.6).

```
> pch <- rep(".", n)

> pch[outliers] <- "+"

> col <- rep("black", n)

> col[outliers] <- "red"

> pairs(iris2, pch=pch, col=col)
```

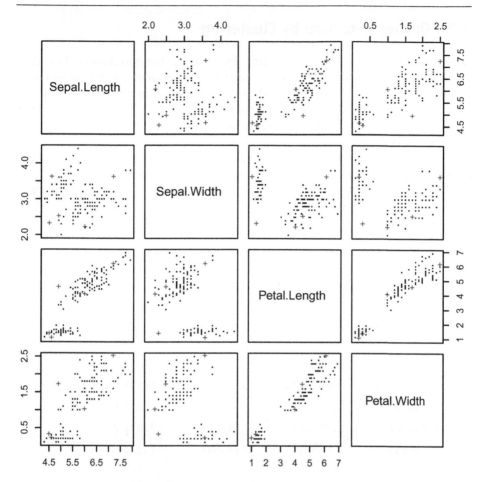

Figure 7.6 Outliers in a matrix of scatter plots.

Package *Rlof* (Hu et al., 2011) provides function lof(), a parallel implementation of the LOF algorithm. Its usage is similar to lofactor(), but lof() has two additional features of supporting multiple values of *k* and several choices of distance metrics. Below is an example of lof(). After computing outlier scores, outliers can be detected by selecting the top ones. Note that the current version of package *Rlof* (v1.0.0) works under Mac OS X, but does not work under Windows, because it depends on package *multicore* for parallel computing.

```
> library(Rlof)

> outlier.scores <- lof(iris2, k=5)

> # try with different number of neighbors (k = 5,6,7,8,9 and 10)

> outlier.scores <- lof(iris2, k=c(5:10))
```

7.3 Outlier Detection by Clustering

Another way to detect outliers is clustering. By grouping data into clusters, those data not assigned to any clusters are taken as outliers. For example, with density-based clustering such as DBSCAN (Ester et al., 1996), objects are grouped into one cluster if they are connected to one another by densely populated area. Therefore, objects not assigned to any clusters are isolated from other objects and are taken as outliers. An example of DBSCAN can be found in Section 6.4 Density-based Clustering.

We can also detect outliers with the *k*-means algorithm. With *k*-means, the data are partitioned into *k* groups by assigning them to the closest cluster centers. After that, we can calculate the distance (or dissimilarity) between each object and its cluster center, and pick those with largest distances as outliers. An example of outlier detection with *k*-means from the `iris` data (see Section 1.3.1 for details of the data) is given below.

```
> # remove species from the data to cluster

> iris2 <- iris[,1:4]

> kmeans.result <- kmeans(iris2, centers=3)

> # cluster centers

> kmeans.result$centers

    Sepal.Length    Sepal.Width    Petal.Length    Petal.Width
1   5.006000        3.428000       1.462000        0.246000
2   6.850000        3.073684       5.742105        2.071053
3   5.901613        2.748387       4.393548        1.433871

> # cluster IDs

> kmeans.result$cluster

  [1] 1 1 1 1 1 1 1 1 1 1 1 1 1 1 1 1 1 1 1 1 1 1 1 1 1 1 1 1 1 1 1 1 1 1 1 1 1
 [38] 1 1 1 1 1 1 1 1 1 1 1 1 1 3 3 2 3 3 3 3 3 3 3 3 3 3 3 3 3 3 3 3 3 3 3 3 3
 [75] 3 3 3 2 3 3 3 3 3 3 3 3 3 3 3 3 3 3 3 3 3 3 3 3 3 3 3 2 3 2 2 2 3 2 2 2 2
[112] 2 2 3 3 2 2 2 2 3 2 3 2 3 2 2 3 3 2 2 2 2 2 3 2 2 2 2 3 2 2 2 3 2 2 2 3 2
[149] 2 3

> # calculate distances between objects and cluster centers

> centers <- kmeans.result$centers[kmeans.result$cluster,]

> distances <- sqrt(rowSums((iris2 - centers)^2))

> # pick top 5 largest distances
```

```
> outliers <- order(distances, decreasing=T)[1:5]

> # who are outliers

> print(outliers)

[1] 99 58 94 61 119

> print(iris2[outliers,])
```

	Sepal.Length	Sepal.Width	Petal.Length	Petal.Width
99	5.1	2.5	3.0	1.1
58	4.9	2.4	3.3	1.0
94	5.0	2.3	3.3	1.0
61	5.0	2.0	3.5	1.0
119	7.7	2.6	6.9	2.3

```
> # plot clusters

> plot(iris2[,c("Sepal.Length", "Sepal.Width")], pch="o",

+       col=kmeans.result$cluster, cex=0.3)

> # plot cluster centers

> points(kmeans.result$centers[,c("Sepal.Length",
  "Sepal.Width")], col=1:3, pch=8, cex=1.5)
```

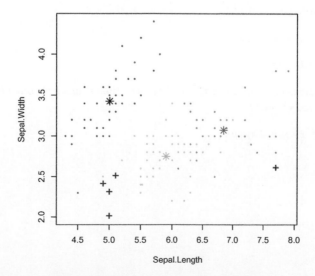

Figure 7.7 Outliers with k-means clustering.

```
> # plot outliers

> points(iris2[outliers, c("Sepal.Length", "Sepal.Width")],
  pch="+", col=4, cex=1.5)
```

In the above figure, cluster centers are labeled with asterisks and outliers with "+" (see Figure 7.7).

7.4   Outlier Detection from Time Series

This section presents an example of outlier detection from time series data. In this example, the time series data are first decomposed with robust regression using function stl() and then outliers are identified. An introduction of STL (Seasonal-trend decomposition based on Loess) (Cleveland et al., 1990) is available at http://cs.wellesley.edu/~cs315/Papers/stl%20statistical%20model.pdf. More examples of time series decomposition can be found in Section 8.2.

```
> # use robust fitting

> f <- stl(AirPassengers, "periodic", robust=TRUE)

>(outliers <- which(f$weights<1e-8))

[1] 79 91 92 102 103 104 114 115 116 126 127 128 138 139 140

> # set layout

> op <- par(mar=c(0, 4, 0, 3), oma=c(5, 0, 4, 0), mfcol=c(4, 1))

> plot(f, set.pars=NULL)

> sts <- f$time.series

> # plot outliers

> points(time(sts)[outliers], 0.8*sts[,"remainder"][outliers],
  pch="x", col="red")

> par(op) # reset layout
```

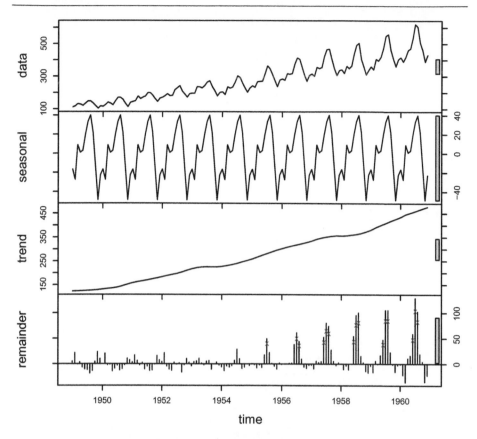

Figure 7.8 Outliers in time series data.

In the above figure, outliers are labeled with "x" in red (see Figure 7.8).

7.5 Discussions

The LOF algorithm is good at detecting local outliers, but it works on numeric data only. Package *Rlof* relies on the *multicore* package, which does not work under Windows. A fast and scalable outlier detection strategy for categorical data is the Attribute Value Frequency (AVF) algorithm (Koufakou et al., 2007).

Some other R packages for outlier detection are:

- Package *extremevalues* (van der Loo, 2010): univariate outlier detection;

- Package *mvoutlier* (Filzmoser and Gschwandtner, 2012): multivariate outlier detection based on robust methods; and

- Package *outliers* (Komsta, 2011): tests for outliers.

8 Time Series Analysis and Mining

This chapter presents examples on time series decomposition, forecasting, clustering, and classification. The first section introduces briefly time series data in R. The second section shows an example on decomposing time series into trend, seasonal, and random components. The third section presents how to build an autoregressive integrated moving average (ARIMA) model in R and use it to predict future values. The fourth section introduces Dynamic Time Warping (DTW) and hierarchical clustering of time series data with Euclidean distance and with DTW distance. The fifth section shows three examples on time series classification: one with original data, the other with DWT (Discrete Wavelet Transform) transformed data, and another with k-NN classification. The chapter ends with discussions and further readings.

8.1 Time Series Data in R

Class `ts` represents data which has been sampled at equispaced points in time. A frequency of seven indicates that a time series is composed of weekly data, and 12 and 4 are used, respectively, for monthly and quarterly series. An example below shows the construction of a time series with 30 values (1–30). `Frequency=12` and `start=c(2011,3)` specify that it is a monthly series starting from March 2011.

```
> a <- ts(1:30, frequency=12, start=c(2011,3))
> print(a)

        Jan  Feb  Mar  Apr  May  Jun  Jul  Aug  Sep  Oct  Nov  Dec
2011               1    2    3    4    5    6    7    8    9   10
2012   11   12   13   14   15   16   17   18   19   20   21   22
2013   23   24   25   26   27   28   29   30
> str(a)
Time-Series [1:30] from 2011 to 2014: 1 2 3 4 5 6 7 8 9 10 …
> attributes(a)

$tsp

[1] 2011.167 2013.583 12.000

$class

[1] "ts"
```

R and Data Mining. DOI: http://dx.doi.org/10.1016/B978-0-12-396963-7.00008-8

RDM

8.2   Time Series Decomposition

Time Series Decomposition is to decompose a time series into trend, seasonal, cyclical, and irregular components. The trend component stands for long-term trend, the seasonal component is seasonal variation, the cyclical component is repeated but non-periodic fluctuations, and the residuals are irregular component.

A time series of AirPassengers is used below as an example to demonstrate time series decomposition. It is composed of monthly totals of Box & Jenkins international airline passengers from 1949 to 1960. It has 144(=12*12) values (see Figure 8.1).

```
> plot(AirPassengers)
```

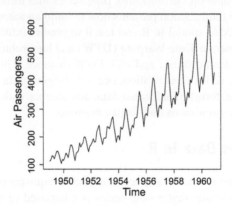

Figure 8.1 A time series of AirPassengers.

Function decompose() is applied below to AirPassengers to break it into various components (see Figures 8.2 and 8.3).

```
> # decompose time series
> apts <- ts(AirPassengers, frequency=12)
> f <- decompose(apts)
> # seasonal figures
> f$figure
[1] -24.748737 -36.188131 -2.241162 -8.036616  -4.506313 35.402778
[7] 63.830808  62.823232  16.520202 -20.642677 -53.593434 -28.619949

> plot(f$figure, type="b", xaxt="n", xlab="")
> # get names of 12 months in English words
> monthNames <- months(ISOdate(2011,1:12,1))
> # label x-axis with month names
> # las is set to 2 for vertical label orientation
> axis(1, at=1:12, labels=monthNames, las=2)
```

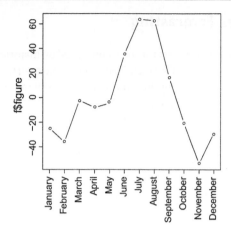

Figure 8.2 Seasonal component.

```
> plot(f)
```

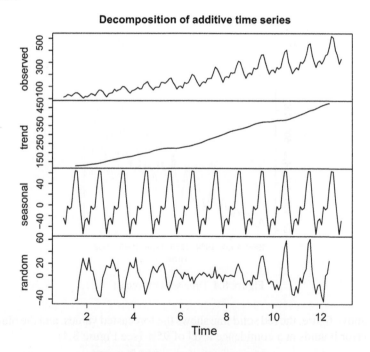

Figure 8.3 Time series decomposition.

In Figure 8.3, the first chart is the original time series. The second is trend of the data, the third shows seasonal factors, and the last chart is the remaining components after removing trend and seasonal factors.

Some other functions for time series decomposition are stl() in package *stats* (R Development Core Team, 2012), decomp() in package *timsac* (The Institute of Statistical Mathematics, 2012), and tsr() in package *ast*.

8.3   Time Series Forecasting

Time series forecasting is to forecast future events based on historical data. One example is to predict the opening price of a stock based on its past performance. Two popular models for time series forecasting are autoregressive moving average (ARMA) and autoregressive integrated moving average (ARIMA).

Here is an example to fit an ARIMA model to a univariate time series and then use it for forecasting.

```
> fit <- arima(AirPassengers, order=c(1,0,0), list(order=c(2,1,0),
  period=12))
> fore <- predict(fit, n.ahead=24)
> # error bounds at 95% confidence level
> U <- fore$pred +2*fore$se
> L <- fore$pred - 2*fore$se
> ts.plot(AirPassengers, fore$pred, U, L, col=c(1,2,4,4),
  lty=c(1,1,2,2))
> legend("topleft", c("Actual", "Forecast",
  "Error Bounds(95% Confidence)"), col=c(1,2,4), lty=c(1,1,2))
```

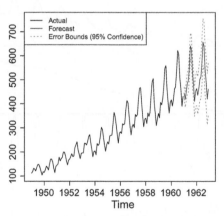

Figure 8.4 Time series forecast.

In the above figure, the red solid line shows the forecasted values, and the blue dotted lines are error bounds at a confidence level of 95% (see Figure 8.4).

8.4   Time Series Clustering

Time series clustering is to partition time series data into groups based on similarity or distance, so that time series in the same cluster are similar to each other. There are various measures of distance or dissimilarity, such as Euclidean distance, Manhattan distance, Maximum norm, Hamming distance, the angle between two vectors (inner product), and Dynamic Time Warping (DTW) distance.

8.4.1 Dynamic Time Warping

Dynamic Time Warping (DTW) finds optimal alignment between two time series (Keogh and Pazzani, 2001) (see Figure 8.5) and an implement of it in R is package *dtw* (Giorgino, 2009). In that package, function dtw (x, y, …) computes dynamic time warp and finds optimal alignment between two time series x and y, and dtwDist(mx, my=mx, …) or dist(mx, my=mx, method="DTW", …) calculates the distances between time series mx and my.

```
> library(dtw)
> idx <- seq(0, 2*pi, len=100)
> a <- sin(idx) + runif(100)/10
> b <- cos(idx)
> align <- dtw(a, b, step=asymmetricP1, keep=T)
> dtwPlotTwoWay(align)
```

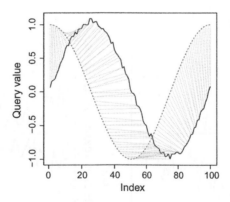

Figure 8.5 Alignment with dynamic time warping.

8.4.2 Synthetic Control Chart Time Series Data

The synthetic control chart time series[1] is used in the examples in the following sections. The dataset contains 600 examples of control charts synthetically generated by the process in Alcock and Manolopoulos (1999). Each control chart is a time series with 60 values, and there are six classes (see Figure 8.6):

- 1–100: Normal;
- 101–200: Cyclic;
- 201–300: Increasing trend;

[1] http://kdd.ics.uci.edu/databases/synthetic_control/synthetic_control.html

- 301–400: Decreasing trend;
- 401–500: Upward shift; and
- 501–600: Downward shift.

Firstly, the data is read into R with read.table(). Parameter sep is set to "" (no space between double quotation marks), which is used when the separator is white space, i.e. one or more spaces, tabs, newlines, or carriage returns.

```
> sc <- read.table("./data/synthetic_control.data", header=F,
    sep="")
> # show one sample from each class
> idx <- c(1,101,201,301,401,501)
> sample1 <- t(sc[idx,])
> plot.ts(sample1, main="")
```

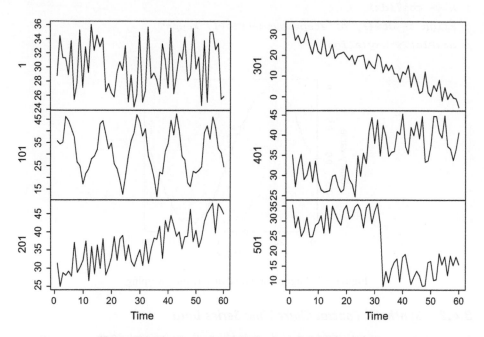

Figure 8.6 Six classes in synthetic control chart time series.

8.4.3   *Hierarchical Clustering with Euclidean Distance*

At first, we select ten cases randomly from each class. Otherwise, there will be too many cases and the plot of hierarchical clustering will be over crowded.

```
> set.seed(6218)
> n <- 10
> s <- sample(1:100, n)
```

```
> idx <- c(s, 100+s, 200+s, 300+s, 400+s, 500+s)
> sample2 <- sc[idx,]
> observedLabels <- rep(1:6, each=n)
> # hierarchical clustering with Euclidean distance
> hc <- hclust(dist(sample2), method="average")
> plot(hc, labels=observedLabels, main="")
> # cut tree to get 6 clusters
> rect.hclust(hc, k=6)
```

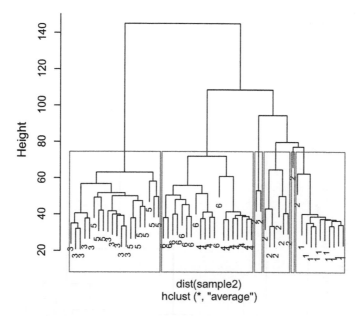

Figure 8.7 Hierarchical clustering with euclidean distance.

```
> memb <- cutree(hc, k=6)
> table(observedLabels, memb)
```

```
                memb
observedLabels  1    2   3   4   5   6
1               10   0   0   0   0   0
2               1    6   2   1   0   0
3               0    0   0   0   10  0
4               0    0   0   0   0   10
5               0    0   0   0   10  0
6               0    0   0   0   0   10
```

The clustering result in Figure 8.7 shows that, increasing trend (class 3) and upward shift (class 5) are not well separated, and decreasing trend (class 4) and downward shift (class 6) are also mixed.

8.4.4   Hierarchical Clustering with DTW Distance

Next, we try hierarchical clustering with the DTW distance.

```
> library(dtw)

> distMatrix <- dist(sample2, method="DTW")

> hc <- hclust(distMatrix, method="average")

> plot(hc, labels=observedLabels, main="")

> # cut tree to get 6 clusters

> rect.hclust(hc, k=6)
```

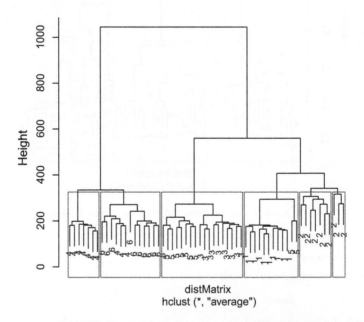

Figure 8.8 Hierarchical clustering with DTW distance.

```
> memb <- cutree(hc, k=6)

> table(observedLabels, memb)
```

	memb					
observedLabels	1	2	3	4	5	6
1	10	0	0	0	0	0
2	0	7	3	0	0	0
3	0	0	0	10	0	0
4	0	0	0	0	7	3
5	2	0	0	8	0	0
6	0	0	0	0	0	10

By comparing Figure 8.8 with Figure 8.7, we can see that the DTW distance are better than the Euclidean distance for measuring the similarity between time series.

8.5   Time Series Classification

Time series classification is to build a classification model based on labeled time series and then use the model to predict the label of unlabeled time series. New features extracted from time series may help to improve the performance of classification models. Techniques for feature extraction include Singular Value Decomposition (SVD), Discrete Fourier Transform (DFT), Discrete Wavelet Transform (DWT), Piecewise Aggregate Approximation (PAA), Perpetually Important Points (PIP), Piecewise Linear Representation, and Symbolic Representation.

8.5.1   Classification with Original Data

We use ctree() from package *party* (Hothorn et al., 2010) to demonstrate classification of time series with the original data. The class labels are changed into categorical values before feeding the data into ctree(), so that we won't get class labels as a real number like 1.35. The built decision tree is shown in Figure 8.9.

```
> classId <- rep(as.character(1:6), each=100)

> newSc <- data.frame(cbind(classId, sc))

> library(party)

> ct <- ctree(classId ~ ., data=newSc,

+            controls =ctree_control(minsplit=30, minbucket=10,
             maxdepth=5))

> pClassId <- predict(ct)
```

```
> table(classId, pClassId)
```

	pClassId					
classId	1	2	3	4	5	6
1	97	0	0	0	0	3
2	1	93	2	0	0	4
3	0	0	96	0	4	0
4	0	0	0	100	0	0
5	4	0	10	0	86	0
6	0	0	0	87	0	13

```
> # accuracy

> (sum(classId==pClassId)) / nrow(sc)

[1] 0.8083333

> plot(ct, ip_args=list(pval=FALSE), ep_args=list(digits=0))
```

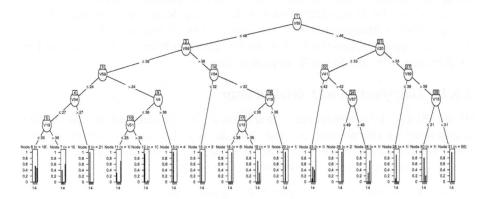

Figure 8.9 Decision tree.

8.5.2 Classification with Extracted Features

Next, we use DWT (Discrete Wavelet Transform) (Burrus et al., 1998) to extract features from time series and then build a classification model. Wavelet transform provides a multi-resolution representation using wavelets. An example of Haar Wavelet Transform, the simplest DWT, is available at http://dmr.ath.cx/gfx/haar/. Another popular feature extraction technique is Discrete Fourier Transform (DFT) (Agrawal et al., 1993).

An example on extracting DWT (with Haar filter) coefficients is shown below. Package *wavelets* (Aldrich, 2010) are used for discrete wavelet transform. In the package, function dwt(X, filter, n.levels, …) computes the discrete wavelet transform coefficients, where X is a univariate or multivariate time series, filter indicates which wavelet filter to use, and n.levels specifies the level of decomposition. It returns an object of class dwt, whose slot W contains wavelet coefficients and V contains scaling coefficients. The original time series can be reconstructed via an inverse discrete wavelet transform with function idwt() in the same package. The produced model is shown in Figure 8.10.

```
> library(wavelets)

> wtData <- NULL

> for(i in 1:nrow(sc)) {

+    a <- t(sc[i,])

+    wt <- dwt(a, filter="haar", boundary="periodic")

+    wtData <- rbind(wtData, unlist(c(wt@W, wt@V[[wt@level]])))

+ }

> wtData <- as.data.frame(wtData)

> wtSc <- data.frame(cbind(classId, wtData))

> # build a decision tree with DWT coefficients

> ct <- ctree(classId ~ ., data=wtSc,

    controls =ctree_control(minsplit=30, minbucket=10, maxdepth=5))

> pClassId <- predict(ct)

> table(classId, pClassId)

          pClassId
classId  1          2    3    4    5    6

1        97         3    0    0    0    0

2        1          99   0    0    0    0

3        0          0    81   0    19   0

4        0          0    0    63   0    37

5        0          0    16   0    84   0

6        0          0    0    1    0    99


>(sum(classId==pClassId))/ nrow(wtSc)

[1] 0.8716667
```

```
> plot(ct, ip_args=list(pval=FALSE), ep_args=list(digits=0))
```

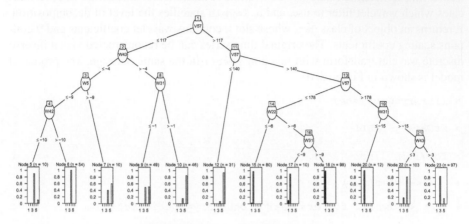

Figure 8.10 Decision tree with DWT.

8.5.3   k-NN Classification

The *k*-NN classification can also be used for time series classification. It finds out the *k* nearest neighbors of a new instance and then labels it by majority voting. However, the time complexity of a naive way to find *k* nearest neighbors is $O(n^2)$, where *n* is the size of data. Therefore, an efficient indexing structure is needed for large datasets. Package *RANN* supports fast nearest neighbor search with a time complexity of $O(n \log n)$ using Arya and Mount's ANN library (v1.1.1).[2] Below is an example of *k*-NN classification of time series without indexing.

```
> k <- 20

> # create a new time series by adding noise to time series 501

> newTS <- sc[501,] + runif(100)*15

> distances <- dist(newTS, sc, method="DTW")

> s <- sort(as.vector(distances), index.return=TRUE)

> # class IDs of k nearest neighbors

> table(classId[s$ix[1:k]])

  4    6

  3   17
```

---

[2] http://www.cs.umd.edu/~mount/ANN/

For the 20 nearest neighbors of the new time series, three of them are of class 4, and 17 are of class 6. With majority voting, that is, taking the more frequent label as winner, the label of the new time series is set to class 6.

8.6   Discussions

There are many R functions and packages available for time series decomposition and forecasting. However, there are no R functions or packages specially for time series classification and clustering. There are a lot of research publications on techniques specially for classifying/clustering time series data, but there are no R implementations for them (as far as I know).

To do time series classification, one is suggested to extract and build features first, and then apply existing classification techniques, such as SVM, k-NN, neural networks, regression, and decision trees, to the feature set.

For time series clustering, one needs to work out his/her own distance or similarity metrics, and then use existing clustering techniques, such as k-means or hierarchical clustering, to find clusters.

8.7   Further Readings

An introduction of R functions and packages for time series is available as *CRAN Task View: Time Series Analysis* at   http://cran.r-project.org/web/views/ TimeSeries.html.

R code examples for time series can be found in slides *Time Series Analysis and Mining with R* at http://www.rdatamining.com/docs.

Some further readings on time series representation, similarity, clustering, and classification are Agrawal et al. (1993), Burrus et al. (1998), Chan and Fu (1999), Chan and Yu (2003), Keogh and Pazzani (1998), Keogh et al. (2000), Keogh and Pazzani (2000), Mörchen (2003), 1998, Vlachos et al. (2003), Wu et al. (2000), Zhao and Zhang (2006).

9 Association Rules

This chapter presents examples of association rule mining with *R*. It starts with basic concepts of association rules, and then demonstrates association rules mining with *R*. After that, it presents examples of pruning redundant rules and interpreting and visualizing association rules. The chapter concludes with discussions and recommended readings.

9.1 Basics of Association Rules

Association rules are rules presenting association or correlation between itemsets. An association rule is in the form of $A \Rightarrow B$, where A and B are two disjoint itemsets, referred to respectively, as the `lhs` (left-hand side) and `rhs` (right-hand side) of the rule. The three most widely used measures for selecting interesting rules are *support*, *confidence*, and *lift*. *Support* is the percentage of cases in the data that contains both A and B, *confidence* is the percentage of cases containing A that also contain B, and *lift* is the ratio of confidence to the percentage of cases containing B. The formulae to calculate them are:

$$\text{support}(A \Rightarrow B) = P(A \cup B), \tag{9.1}$$

$$\text{confidence}(A \Rightarrow B) = P(B|A), \tag{9.2}$$

$$= \frac{P(A \cup B)}{P(A)}, \tag{9.3}$$

$$\text{lift}(A \Rightarrow B) = \frac{\text{confidence}(A \Rightarrow B)}{P(B)}, \tag{9.4}$$

$$= \frac{P(A \cup B)}{P(A)P(B)}, \tag{9.5}$$

where $P(A)$ is the percentage (or probability) of cases containing A.

In addition to support, confidence, and lift, there are many other interestingness measures, such as chi-square, conviction, gini, and leverage. An introduction to over 20 measures can be found in Tan et al.'s work (Tan et al., 2002).

R and Data Mining. DOI: http://dx.doi.org/10.1016/B978-0-12-396963-7.00009-X

9.2   The Titanic Dataset

The Titanic dataset in the *datasets* package is a four-dimensional table with sum-
marized information on the fate of passengers on the Titanic according to social
class, sex, age, and survival. To make it suitable for association rule mining, we
reconstruct the raw data as titanic.raw, where each row represents a person.
The reconstructed raw data can also be downloaded as file "titanic.raw.rdata" at
http://www.rdatamining.com/data.

```
> str(Titanic)

  table [1:4, 1:2, 1:2, 1:2] 0 0 35 0 0 0 17 0 118 154 ...

  — attr(*, "dimnames")=List of 4

    ..$ Class    : chr [1:4] "1st" "2nd" "3rd" "Crew"

    ..$ Sex      : chr [1:2] "Male" "Female"

    ..$ Age      : chr [1:2] "Child" "Adult"

    ..$ Survived: chr [1:2] "No" "Yes"

> df <- as.data.frame(Titanic)

> head(df)
```

	Class	Sex	Age	Survived	Freq
1	1st	Male	Child	No	0
2	2nd	Male	Child	No	0
3	3rd	Male	Child	No	35
4	Crew	Male	Child	No	0
5	1st	Female	Child	No	0
6	2nd	Female	Child	No	0

```
> titanic.raw <- NULL

> for(i in 1:4) {

+    titanic.raw <- cbind(titanic.raw, rep(as.character(df[,i]),
  df$Freq))

+}

> titanic.raw <- as.data.frame(titanic.raw)

> names(titanic.raw) <- names(df)[1:4]
```

```
> dim(titanic.raw)

[1] 2201 4

> str(titanic.raw)


'data.frame' 2201 obs. of 4 variables:
$ Class   : Factor w/ 4 levels "1st","2nd","3rd",..: 3 3 3 3 3 3 3 3 3 3 …
$ Sex     : Factor w/ 2 levels "Female","Male": 2 2 2 2 2 2 2 2 2 2 …
$ Age     : Factor w/ 2 levels "Adult","Child": 2 2 2 2 2 2 2 2 2 2 …
$ Survived: Factor w/ 2 levels "No","Yes": 1 1 1 1 1 1 1 1 1 1 …


> head(titanic.raw)


    Class  Sex    Age    Survived
1   3rd    Male   Child  No
2   3rd    Male   Child  No
3   3rd    Male   Child  No
4   3rd    Male   Child  No
5   3rd    Male   Child  No
6   3rd    Male   Child  No


> summary(titanic.raw)


Class       Sex          Age          Survived
1st :325    Female:470   Adult:2092   No :1490
2nd :285    Male  :1731  Child:109    Yes:711
3rd :706
Crew:885
```

Now we have a dataset where each row stands for a person, and it can be used for association rule mining.

The raw Titanic dataset can also be downloaded from http://www.cs.toronto. edu/delve/data/titanic/desc.html. The data is file "Dataset.data" in the compressed archive "titanic.tar.gz". It can be read into *R* with the code below.

```
> # have a look at the 1st 5 lines

> readLines("./data/Dataset.data", n=5)

[1] "1st adult male yes" "1st adult male yes" "1st adult male
  yes"

[4] "1st adult male yes" "1st adult male yes"

> # read it into R

> titanic <- read.table("./data/Dataset.data", header=F)

> names(titanic) <- c("Class", "Sex", "Age", "Survived")
```

9.3 Association Rule Mining

A classic algorithm for association rule mining is APRIORI (Agrawal and Srikant, 1994). It is a level-wise, breadth-first algorithm which counts transactions to find frequent itemsets and then derive association rules from them. An implementation of it is function apriori() in package *arules* (Hahsler et al., 2011). Another algorithm for association rule mining is the ECLAT algorithm (Zaki, 2000), which finds frequent itemsets with equivalence classes, depth-first search and set intersection instead of counting. It is implemented as function eclat() in the same package.

Below we demonstrate association rule mining with apriori(). With the function, the default settings are: (1) supp=0.1, which is the minimum support of rules; (2) conf=0.8, which is the minimum confidence of rules; and (3) maxlen=10, which is the maximum length of rules.

```
> library(arules)

> # find association rules with default settings

> rules.all <- apriori(titanic.raw)

parameter specification:

  confidence minval smax  arem aval  originalSupport support minlen maxlen target

  0.8         0.1    1     none FALSE TRUE            0.1     1      10     rules

  ext

FALSE

  algorithmic control:

  filter  tree  heap   memopt  load  sort  verbose

  0.1     TRUE  TRUE   FALSE   TRUE  2     TRUE
```

```
apriori - find association rules with the apriori algorithm
version 4.21 (2004.05.09)     (c) 1996-2004 Christian Borgelt
set item appearances … [0 item(s)] done [0.00s].
set transactions … [10 item(s), 2201 transaction(s)] done [0.00s].
sorting and recoding items … [9 item(s)] done [0.00s].
creating transaction tree … done [0.00s].
checking subsets of size 1 2 3 4 done [0.00s].
writing … [27 rule(s)] done [0.00s].
creating S4 object … done [0.00s].
> rules.all
set of 27 rules
> inspect(rules.all)

   lhs                rhs              support   confidence lift
1  {}              => {Age=Adult}    0.9504771 0.9504771  1.0000000
2  {Class=2nd}     => {Age=Adult}    0.1185825 0.9157895  0.9635051
3  {Class=1st}     => {Age=Adult}    0.1449341 0.9815385  1.0326798
4  {Sex=Female}    => {Age=Adult}    0.1930940 0.9042553  0.9513700
5  {Class=3rd}     => {Age=Adult}    0.2848705 0.8881020  0.9343750
6  {Survived=Yes}  => {Age=Adult}    0.2971377 0.9198312  0.9677574
7  {Class=Crew}    => {Sex=Male}     0.3916402 0.9740113  1.2384742
8  {Class=Crew}    => {Age=Adult}    0.4020900 1.0000000  1.0521033
9  {Survived=No}   => {Sex=Male}     0.6197183 0.9154362  1.1639949
10 {Survived=No}   => {Age=Adult}    0.6533394 0.9651007  1.0153856
11 {Sex=Male}      => {Age=Adult}    0.7573830 0.9630272  1.0132040
12 {Sex=Female,
    Survived=Yes} => {Age=Adult}    0.1435711 0.9186047  0.9664669
13 {Class=3rd,
    Sex=Male}     => {Survived=No}  0.1917310 0.8274510  1.2222950
```

```
14 {Class=3rd,
     Survived=No}  => {Age=Adult}    0.2162653 0.9015152 0.9484870
15 {Class=3rd,
     Sex=Male}     => {Age=Adult}    0.2099046 0.9058824 0.9530818
16 {Sex=Male,
     Survived=Yes} => {Age=Adult}    0.1535666 0.9209809 0.9689670
17 {Class=Crew,
     Survived=No}  => {Sex=Male}     0.3044071 0.9955423 1.2658514
18 {Class=Crew,
     Survived=No}  => {Age=Adult}    0.3057701 1.0000000 1.0521033
19 {Class=Crew,
     Sex=Male}     => {Age=Adult}    0.3916402 1.0000000 1.0521033
20 {Class=Crew,
     Age=Adult}    => {Sex=Male}     0.3916402 0.9740113 1.2384742
21 {Sex=Male,
     Survived=No}  => {Age=Adult}    0.6038164 0.9743402 1.0251065
22 {Age=Adult,
     Survived=No}  => {Sex=Male}     0.6038164 0.9242003 1.1751385
23 {Class=3rd,
     Sex=Male,
     Survived=No}  => {Age=Adult}    0.1758292 0.9170616 0.9648435
24 {Class=3rd,
     Age=Adult,
     Survived=No}  => {Sex=Male}     0.1758292 0.8130252 1.0337773
25 {Class=3rd,
     Sex=Male,
     Age=Adult}    => {Survived=No}  0.1758292 0.8376623 1.2373791
26 {Class=Crew,
     Sex=Male,
     Survived=No}  => {Age=Adult}    0.3044071 1.0000000 1.0521033
```

```
27 {Class=Crew,

     Age=Adult,

     Survived=No} => {Sex=Male} 0.3044071 0.9955423 1.2658514
```

As a common phenomenon for association rule mining, many rules generated above are uninteresting. Suppose that we are interested in only rules with rhs indicating survival, so we set rhs=c("Survived=No", "Survived=Yes") in appearance to make sure that only "Survived=No" and "Survived=Yes" will appear in the rhs of rules. All other items can appear in the lhs, as set with default="lhs". In the above result rules.all, we can also see that the left-hand side (lhs) of the first rule is empty. To exclude such rules, we set minlen to 2 in the code below. Moreover, the details of progress are suppressed with verbose=F. After association rule mining, rules are sorted by lift to make high-lift rules appear first.

```
> # rules with rhs containing "Survived" only

> rules <- apriori(titanic.raw, control = list(verbose=F),

+           parameter = list(minlen=2, supp=0.005, conf=0.8),

+           appearance = list(rhs=c("Survived=No", "Survived=Yes"),

+                default="lhs"))

> quality(rules) <- round(quality(rules), digits=3)

> rules.sorted <- sort(rules, by="lift")

> inspect(rules.sorted)
```

	lhs	rhs	support	confidence	lift
1	{Class=2nd,				
	Age=Child}	=> {Survived=Yes}	0.011	1.000	3.096
2	{Class=2nd,				
	Sex=Female,				
	Age=Child}	=> {Survived=Yes}	0.006	1.000	3.096
3	{Class=1st,				
	Sex=Female}	=> {Survived=Yes}	0.064	0.972	3.010
4	{Class=1st,				
	Sex=Female,				
	Age=Adult}	=> {Survived=Yes}	0.064	0.972	3.010

```
5    {Class=2nd,

     Sex=Female}    => {Survived=Yes}   0.042   0.877   2.716

6    {Class=Crew,

     Sex=Female}    => {Survived=Yes}   0.009   0.870   2.692

7    {Class=Crew,

     Sex=Female,

     Age=Adult}     => {Survived=Yes}   0.009   0.870   2.692

8    {Class=2nd,

     Sex=Female,

     Age=Adult}     => {Survived=Yes}   0.036   0.860   2.663

9    {Class=2nd,

     Sex=Male,

     Age=Adult}     => {Survived=No}    0.070   0.917   1.354

10   {Class=2nd,

     Sex=Male}      => {Survived=No}    0.070   0.860   1.271

11   {Class=3rd,

     Sex=Male,

     Age=Adult}     => {Survived=No}    0.176   0.838   1.237

12   {Class=3rd,

     Sex=Male}      => {Survived=No}    0.192   0.827   1.222
```

When other settings are unchanged, with a lower minimum support, more rules will be produced, and the associations between itemsets shown in the rules will be more likely to be by chance. In the above code, the minimum support is set to 0.005, so each rule is supported at least by 12(=ceiling (0.005 * 2201)) cases, which is acceptable for a population of 2201.

Support, confidence, and lift are three common measures for selecting interesting association rules. Besides them, there are many other interestingness measures, such as chi-square, conviction, gini, and leverage (Tan et al., 2002). More than 20 measures can be calculated with function interestMeasure() in the *arules* package.

9.4   Removing Redundancy

Some rules generated in the previous section (see rules.sorted, p. 95) provide little or no extra information when some other rules are in the result. For example, the above rule 2 provides no extra knowledge in addition to rule 1, since rule 1 tells us that all 2nd-class children survived. Generally speaking, when a rule (such as rule 2) is a super rule of another rule (such as rule 1) and the former has the same or a lower lift, the former rule (rule 2) is considered to be redundant. Other redundant rules in the above result are rules 4, 7, and 8, compared, respectively, with rules 3, 6, and 5.

Below we prune redundant rules. Note that the rules have already been sorted descendingly by lift.

```
> # find redundant rules

> subset.matrix <- is.subset(rules.sorted, rules.sorted)

> subset.matrix[lower.tri(subset.matrix, diag=T)] <- NA

> redundant <- colSums(subset.matrix, na.rm=T) >= 1

> which(redundant)

[1] 2 4 7 8

> # remove redundant rules

> rules.pruned <- rules.sorted[!redundant]

> inspect(rules.pruned)
```

	lhs	rhs	support	confidence	lift
1	{Class=2nd, Age=Child}	=> {Survived=Yes}	0.011	1.000	3.096
2	{Class=1st, Sex=Female}	=> {Survived=Yes}	0.064	0.972	3.010
3	{Class=2nd, Sex=Female}	=> {Survived=Yes}	0.042	0.877	2.716
4	{Class=Crew, Sex=Female}	=> {Survived=Yes}	0.009	0.870	2.692
5	{Class=2nd, Sex=Male, Age=Adult}	=> {Survived=No}	0.070	0.917	1.354
6	{Class=2nd, Sex=Male}	=> {Survived=No}	0.070	0.860	1.271

```
7  {Class=3rd,

   Sex=Male,

   Age=Adult}   => {Survived=No}   0.176   0.838   1.237

8  {Class=3rd,

   Sex=Male}    => {Survived=No}   0.192   0.827   1.222
```

In the code above, function is.subset(r1, r2) checks whether r1 is a subset of r2 (i.e. whether r2 is a superset of r1). Function lower.tri() returns a logical matrix with TURE in lower triangle. From the above results, we can see that rules 2, 4, 7, and 8 (before redundancy removal) are successfully pruned.

9.5 Interpreting Rules

While it is easy to find high-lift rules from data, it is not an easy job to understand the identified rules. It is not uncommon that the association rules are misinterpreted to find their business meanings. For instance, in the above rule list rules.pruned, the first rule "{Class=2nd, Age=Child} => {Survived=Yes}" has a confidence of one and a lift of three and there are no rules on children of the 1st or 3rd classes. Therefore, it might be interpreted by users as *children of the 2nd class had a higher survival rate than other children*. This is wrong! The rule states only that all children of class 2 survived, but provides no information at all to compare the survival rates of different classes. To investigate the above issue, we run the code below to find rules whose rhs is "Survived=Yes" and lhs contains "Class=1st", "Class=2nd", "Class=3rd", "Age=Child", and "Age=Adult" only, and which contains no other items (default="none"). We use lower thresholds for both support and confidence than before to find all rules for children of different classes.

```
> rules <- apriori(titanic.raw,

+              parameter = list(minlen=3, supp=0.002, conf=0.2),

+              appearance = list(rhs=c("Survived=Yes"),

+                                lhs=c("Class=1st", "Class=2nd",
                                      "Class=3rd",

+                                     "Age=Child", "Age=Adult"),

+                                default="none"),

+              control = list(verbose=F))
```

```
> rules.sorted <- sort(rules, by="confidence")

> inspect(rules.sorted)

   lhs             rhs                  support      confidence lift
1 {Class=2nd,
    Age=Child} => {Survived=Yes} 0.010904134 1.0000000   3.0956399
2 {Class=1st,
    Age=Child} => {Survived=Yes} 0.002726034 1.0000000   3.0956399
3 {Class=1st,
    Age=Adult} => {Survived=Yes} 0.089504771 0.6175549   1.9117275
4 {Class=2nd,
    Age=Adult} => {Survived=Yes} 0.042707860 0.3601533   1.1149048
5 {Class=3rd,
    Age=Child} => {Survived=Yes} 0.012267151 0.3417722   1.0580035
6 {Class=3rd,
    Age=Adult} => {Survived=Yes} 0.068605179 0.2408293   0.7455209
```

In the above result, the first two rules show that children of the 1st class are of the same survival rate as children of the 2nd class and that all of them survived. The rule of 1st-class children did not appear before, simply because its support was below the threshold specified in Section 9.3. Rule 5 presents a sad fact that children of class 3 had a low survival rate of 34%, which is comparable with that of 2nd-class adults (see rule 4) and much lower than 1st-class adults (see rule 3).

9.6   Visualizing Association Rules

Next we show some ways to visualize association rules, including scatter plot, balloon plot, graph, and parallel coordinates plot. More examples on visualizing association rules can be found in the vignettes of package *arulesViz* (Hahsler and Chelluboina, 2012) on CRAN at `http://cran.r-project.org/web/packages/arulesViz/` `vignettes/arulesViz.pdf`.

```
> library(arulesViz)
> plot(rules.all) (see Figure 9.1)
> plot(rules.all, method="grouped") (see Figure 9.2)
```

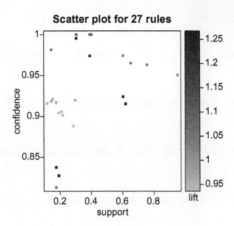

Figure 9.1 A scatter plot of association rules.

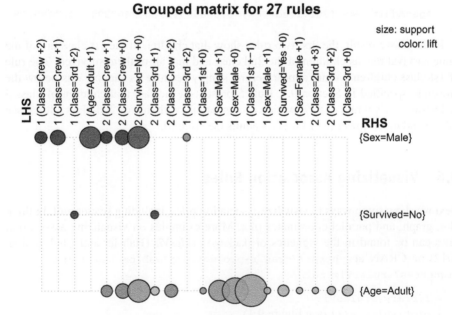

Figure 9.2 A balloon plot of association rules.

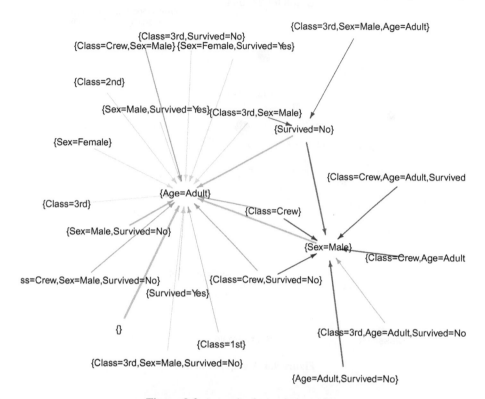

Graph for 27 rules

width: support (0.119 – 0.95)
color: lift (0.934 – 1.266)

Figure 9.3 A graph of association rules.

```
> plot(rules.all, method="graph") (see Figure 9.3)
> plot(rules.all, method="graph", control=list(type="items")) (see
Figure 9.4)
> plot(rules.all, method="paracoord", control=list(reorder=TRUE))
(see Figure 9.5)
```

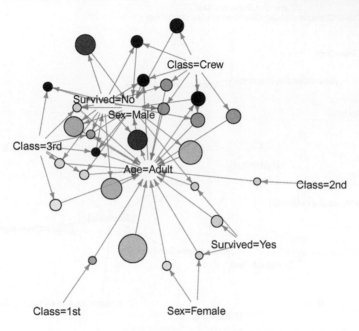

Figure 9.4 A graph of items.

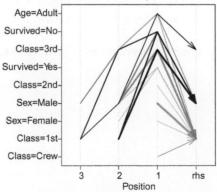

Figure 9.5 A parallel coordinates plot of association rules.

9.7 Discussions and Further Readings

In this chapter, we have demonstrated association rule mining with package *arules* (Hahsler et al., 2011). More examples on that package can be found in Hahsler et al.'s work (Hahsler et al., 2005). Two other packages related to association rules are *arulesSequences* and *arulesNBMiner*. Package *arulesSequences* provides functions for mining sequential patterns (Buchta et al., 2012). Package *arulesNBMiner* implements an algorithm for mining negative binomial (NB) frequent itemsets and NB-precise rules (Hahsler, 2012).

More techniques on post-mining of association rules, such as selecting interesting association rules, visualization of association rules, and using association rules for classification, can be found in Zhao et al.'s work (Zhao, 2009b).

10 Text Mining

This chapter presents examples of text mining with R. Twitter[1] text of @RDataMining is used as the data to analyze. It starts with extracting text from Twitter. The extracted text is then transformed to build a document-term matrix. After that, frequent words and associations are found from the matrix. A word cloud is used to present important words in documents. In the end, words and tweets are clustered to find groups of words and also groups of tweets. In this chapter, "tweet" and "document" will be used interchangeably, so are "word" and "term."

There are three important packages used in the examples: *twitteR*, *tm*, and *wordcloud*. Package *twitteR* (Gentry, 2012) provides access to Twitter data, *tm* (Feinerer, 2012) provides functions for text mining, and *wordcloud* (Fellows, 2012) visualizes the result with a word cloud.[2]

10.1 Retrieving Text from Twitter

Twitter text is used in this chapter to demonstrate text mining. Tweets are extracted from Twitter with the code below using `userTimeline()` in package *twitteR* (Gentry, 2012). Package *twitteR* depends on package *RCurl* (Lang, 2012a), which is available at `http://www.stats.ox.ac.uk/pub/RWin/bin/windows/contrib/`. Another way to retrieve text from Twitter is using package *XML* (Lang, 2012b), and an example of that is given at `http://heuristically.wordpress.com/2011/04/08/text-data-mining-twitter-r/`.

For readers who have no access to Twitter, the tweets data can be downloaded as file "rdmTweets.RData" at `http://www.rdatamining.com/data`. Then readers can skip this section and proceed directly to Section 10.2.

```
> library(twitteR)

> # retrieve the first 200 tweets (or all tweets if fewer than
  200) from the user timeline of @rdatammining

> rdmTweets <- userTimeline("rdatamining", n=200)

> (nDocs <- length(rdmTweets))
```

---

[1] `http://www.twitter.com`.
[2] `http://en.wikipedia.org/wiki/Word_cloud`.

R and Data Mining. DOI: http://dx.doi.org/10.1016/B978-0-12-396963-7.00010-6
© 2013 Yanchang Zhao. Published by Elsevier Inc. All rights reserved.

```
[1] 154
```

Next, we have a look at the five tweets numbered 11 to 15.

```
> rdmTweets[11:15]
```

With the above code, each tweet is printed in one single line, which may exceed the boundary of paper. Therefore, the following code is used in this book to print the five tweets by wrapping the text to fit the width of paper. The same method is used to print tweets in other codes in this chapter.

```
> for (i in 11:15) {
+    cat(paste ("[[", i, "]] ", sep=""))
+    writeLines(strwrap(rdmTweets[[i]]$getText(), width=73))
+ }
```

```
[[11]] Slides on massive data, shared and distributed
    memory,and concurrent programming: bigmemory and foreach
    http://t.co/a6bQzxj5

[[12]] The R Reference Card for Data Mining is updated with
    functions & packages for handling big data & parallel computing.
    http://t.co/FHoVZCyk

[[13]] Post-doc on Optimizing a Cloud for Data Mining primitives,
    INRIA, France  http://t.co/cA28STPO

[[14]] Chief Scientist - Data Intensive Analytics, Pacific
    Northwest National Laboratory (PNNL), US  http://t.co/0Gdzq1Nt
    http://t.co/0Gdzq1Nt

[[15]] Top 10 in Data Mining  http://t.co/7kAuNvuf
```

10.2 Transforming Text

The tweets are first converted to a data frame and then to a corpus, which is a collection of text documents. After that, the corpus can be processed with functions provided in package *tm* (Feinerer, 2012).

```
> # convert tweets to a data frame

> df <- do.call("rbind", lapply(rdmTweets, as.data.frame))

> dim(df)

[1] 154 10
```

```
> library(tm)

> # build a corpus, and specify the source to be character
  vectors

> myCorpus <- Corpus(VectorSource(df$text))
```

After that, the corpus needs a couple of transformations, including changing letters to lower case, and removing punctuations, numbers, and stop words. The general English stop-word list is tailored here by adding "available" and "via" and removing "r" and "big" (for big data). Hyperlinks are also removed in the example below.

```
> # convert to lower case

> myCorpus <- tm_map(myCorpus, tolower)

> # remove punctuation

> myCorpus <- tm_map(myCorpus, removePunctuation)

> # remove numbers

> myCorpus <- tm_map(myCorpus, removeNumbers)

> # remove URLs

> removeURL <- function(x) gsub("http[[:alnum:]]*", "", x)

> myCorpus <- tm_map(myCorpus, removeURL)

> # add two extra stop words: "available" and "via"

> myStopwords <- c(stopwords('english'), "available", "via")

> # remove "r" and "big" from stopwords

> myStopwords <- setdiff(myStopwords, c("r", "big"))

> # remove stopwords from corpus

> myCorpus <- tm_map(myCorpus, removeWords, myStopwords)
```

In the above code, tm_map() is an interface to apply transformations (mappings) to corpora. A list of available transformations can be obtained with getTransformations(), and the mostly used ones are as.PlainTextDocument(), removeNumbers(), removePunctuation(), removeWords(), stemDocument(), and stripWhitespace(). A function removeURL() is defined above to remove hyperlinks, where pattern "http[[:alnum:]]*" matches strings starting with "http" and then followed by any number of alphabetic characters and digits. Strings matching this pattern are removed with gsub(). The above pattern is specified as an regular expression, and details about that can be found by running ?regex in R.

10.3   Stemming Words

In many applications, words need to be stemmed to retrieve their radicals, so that various forms derived from a stem would be taken as the same when counting word frequency. For instance, words "update", "updated", and "updating" would all be stemmed to "updat". Word stemming can be done with the snowball stemmer, which requires packages *Snowball*, *RWeka*, *rJava*, and *RWekajars*. After that, we can complete the stems to their original forms, i.e. "update" for the above example, so that the words would look normal. This can be achieved with function stemCompletion().

```
> # keep a copy of corpus to use later as a dictionary for stem
  completion

> myCorpusCopy <- myCorpus

> # stem words

> myCorpus <- tm_map(myCorpus, stemDocument)

> # inspect documents (tweets) numbered 11 to 15

> # inspect(myCorpus[11:15])

> # The code below is used for to make text fit for paper width

> for (i in 11:15) {

+   cat(paste("[[", i, "]] ", sep=""))

+   writeLines(strwrap(myCorpus[[i]], width=73))

+ }

[[11]] slide massiv data share distribut memoryand concurr
  program bigmemori foreach

[[12]] r refer card data mine updat function packag handl big
  data parallel comput

[[13]] postdoc optim cloud data mine primit inria franc

[[14]] chief scientist data intens analyt pacif northwest nation
  laboratori pnnl

[[15]] top data mine
```

After that, we use stemCompletion() to complete the stems with the unstemmed corpus myCorpusCopy as a dictionary. With the default setting, it takes the most frequent match in dictionary as completion.

```
> # stem completion

> myCorpus <- tm_map(myCorpus, stemCompletion,
  dictionary=myCorpusCopy)
```

Then we have a look at the documents numbered 11 to 15 in the built corpus.

```
> inspect(myCorpus[11:15])
```

```
[[11]] slides massive data share distributed memoryand concurrent
programming foreach

[[12]] r reference card data miners updated functions package
handling big data parallel computing

[[13]] postdoctoral optimizing cloud data miners primitives inria
france

[[14]] chief scientist data intensive analytics pacific northwest
national pnnl

[[15]] top data miners
```

As we can see from the above results, there are something unexpected in the above stemming and completion.

1. In both the stemmed corpus and the completed one, "memoryand" is derived from "...memory, and ..." in the original tweet 11.

2. In tweet 11, word "bigmemory" is stemmed to "bigmemori", and then is removed during stem completion.

3. Word "mining" in tweets 12, 13, & 15 is first stemmed to "mine" and then completed to "miners".

4. "Laboratory" in tweet 14 is stemmed to "laboratori" and then also disappears after completion.

In the above issues, point 1 is caused by the missing of a space after the comma. It can be easily fixed by replacing comma with space before removing punctuation marks in Section 10.2. For points 2 & 4, we have not figured out why it happened like that. Fortunately, the words involved in points 1, 2, & 4 are not important in @RDataMining tweets and ignoring them would not bring any harm to this demonstration of text mining.

Below we focus on point 3, where word "mining" is first stemmed to "mine" and then completed to "miners", instead of "mining", although there are many instances of "mining" in the tweets, compared to only two instances of "miners". There might be a solution for the above problem by changing the parameters and/or dictionaries for stemming and completion, but we failed to find one due to limitation of time and efforts. Instead, we chose a simple way to get around of that by replacing "miners" with "mining", since the latter has many more cases than the former in the corpus. The code for the replacement is given below.

```
> # count frequency of "mining"

> miningCases <- tm_map(myCorpusCopy, grep, pattern="\\<mining")

> sum(unlist(miningCases))

[1] 47

> # count frequency of "miners"

> minerCases <- tm_map(myCorpusCopy, grep, pattern="\\<miners")

> sum(unlist(minerCases))

[1] 2

> # replace "miners" with "mining"

> myCorpus <- tm_map(myCorpus, gsub, pattern="miners",
  replacement="mining")
```

In the first call of function tm_map() in the above code, grep() is applied to every document (tweet) with argument "pattern="\\<mining"". The pattern matches words starting with "mining", where "\<" matches the empty string at the beginning of a word. This ensures that text "rdatamining" would not contribute to the above counting of "mining".

10.4   Building a Term-Document Matrix

A term-document matrix represents the relationship between terms and documents, where each row stands for a term and each column for a document, and an entry is the number of occurrences of the term in the document. Alternatively, one can also build a document-term matrix by swapping row and column. In this section, we build a term-document matrix from the above processed corpus with function TermDocumentMatrix(). With its default setting, terms with less than three characters are discarded. To keep "r" in the matrix, we set the range of wordLengths in the example below.

```
> myTdm <- TermDocumentMatrix(myCorpus,
  control=list(wordLengths=c(1,Inf)))

> myTdm

A term-document matrix (444 terms, 154 documents)

 Non-/sparse entries :   1085/67291

 Sparsity            :   98%

 Maximal term length :   27

 Weighting           :   term frequency (tf)
```

As we can see from the above result, the term-document matrix is composed of 444 terms and 154 documents. It is very sparse, with 98% of the entries being zero. We then have a look at the first six terms starting with "r" and tweets numbered 101 to 110.

```
> idx <- which(dimnames(myTdm)$Terms == "r")
```

```
> inspect(myTdm[idx+(0:5),101:110])
```

```
A term-document matrix (6 terms, 10 documents)

 Non-/sparse entries :   9/51

 Sparsity            :   85%

 Maximal term length :   12

 Weighting           :   term frequency (tf)
```

	Docs									
Terms	101	102	103	104	105	106	107	108	109	110
r	1	1	0	0	2	0	0	1	1	1
ramachandran	0	0	0	0	0	0	0	0	0	0
random	0	0	0	0	0	0	0	0	0	0
ranked	0	0	0	0	0	0	0	0	1	0
rapidminer	1	0	0	0	0	0	0	0	0	0
rdatamining	0	0	0	0	0	0	0	1	0	0

Note that the parameter to control word length used to be `minWordLength` prior to version 0.5-7 of package *tm*. The code to set the minimum word length for old versions of *tm* is below.

```
> myTdm <- TermDocumentMatrix(myCorpus,
  control=list(minWordLength=1))
```

The list of terms can be retrieved with `rownames(myTdm)`. Based on the above matrix, many data mining tasks can be done, for example, clustering, classification and association analysis.

When there are too many terms, the size of a term-document matrix can be reduced by selecting terms that appear in a minimum number of documents, or filtering terms with TF-IDF (term frequency-inverse document frequency) (Wu et al., 2008).

10.5   Frequent Terms and Associations

We have a look at the popular words and the association between words. Note that there are 154 tweets in total.

```
> # inspect frequent words

> findFreqTerms(myTdm, lowfreq=10)

  [1]  "analysis"  "computing"  "data"      "examples"   "introduction"
  [6]  "mining"    "network"    "package"   "positions"  "postdoctoral"
 [11]  "r"         "research"   "slides"    "social"     "tutorial"
 [16]  "users"
```

In the code above, findFreqTerms() finds frequent terms with frequency no less than ten. Note that they are ordered alphabetically, instead of by frequency or popularity.

To show the top frequent words visually, we next make a barplot for them. From the term-document matrix, we can derive the frequency of terms with rowSums(). Then we select terms that appears in ten or more documents and shown them with a barplot using package *ggplot2* (Wickham, 2009). In the code below, geom="bar" specifies a barplot and coord_flip() swaps x- and y-axis. The barplot in Figure 10.1 clearly shows that the three most frequent words are "r," "data" and "mining."

```
> termFrequency <- rowSums(as.matrix(myTdm))

> termFrequency <- subset(termFrequency, termFrequency>=10)

> library(ggplot2)

> qplot(names(termFrequency), termFrequency, geom="bar",
  xlab="Terms") + coord_flip()
```

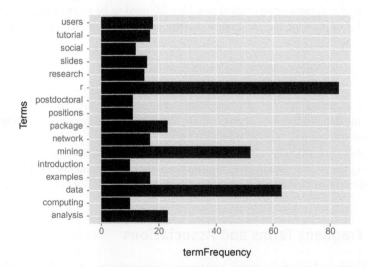

Figure 10.1 Frequent terms.

Alternatively, the above plot can also be drawn with `barplot()` as below, where `las` sets the direction of x-axis labels to be vertical.

```
> barplot(termFrequency, las=2)
```

We can also find what are highly associated with a word with function `findAssocs()`. Below we try to find terms associated with "r" (or "mining") with correlation no less than 0.25, and the words are ordered by their correlation with "r" (or "mining").

```
> # which words are associated with "r"?

> findAssocs(myTdm, 'r', 0.25)

r       users    canberra    cran    list    examples

1.00    0.32     0.26        0.26    0.26    0.25


> # which words are associated with "mining"?

> findAssocs(myTdm, 'mining', 0.25)

mining      data       mahout    recommendation    sets

1.00        0.55       0.39      0.39              0.39

supports    frequent   itemset   card              functions

0.39        0.35       0.34      0.29              0.29

reference   text

0.29        0.26
```

10.6   Word Cloud

After building a term-document matrix, we can show the importance of words with a word cloud (also known as a tag cloud), which can be easily produced with package *wordcloud* (Fellows, 2012). In the code below, we first convert the term-document matrix to a normal matrix, and then calculate word frequencies. After that, we use `wordcloud()` to make a plot for it. With `wordcloud()`, the first two parameters give a list of words and their frequencies. Words with frequency below three are not plotted, as specified by `min.freq=3`. By setting `random.order=F`, frequent words are plotted first, which makes them appear in the center of cloud. We also set the colors to gray levels based on frequency. A colorful cloud can be generated by setting colors with `rainbow()`.

```
> library(wordcloud)

> m <- as.matrix(myTdm)

> # calculate the frequency of words and sort it descendingly by
  frequency
```

```
> wordFreq <- sort(rowSums(m), decreasing=TRUE)

> # word cloud

> set.seed (375) # to make it reproducible

> grayLevels <- gray( (wordFreq+10) / (max(wordFreq)+10))

> wordcloud(words=names(wordFreq), freq=wordFreq, min.freq=3,
  random.order=F, colors=grayLevels)
```

Figure 10.2 Word cloud.

The word cloud in Figure 10.2 clearly shows again that "r," "data" and "mining" are the top three words, which validates that the @RDataMining tweets present information on R and data mining. Some other important words are "analysis," "examples," "slides," "tutorial" and "package," which shows that it focuses on documents and examples on analysis and R packages. Another set of frequent words, "research," "postdoctoral" and "positions," are from tweets about vacancies on post-doctoral and research positions. There are also some tweets on the topic of social network analysis, as indicated by words "network" and "social" in the cloud.

10.7 Clustering Words

We then try to find clusters of words with hierarchical clustering. Sparse terms are removed, so that the plot of clustering will not be crowded with words. Then the distances between terms are calculated with `dist()` after scaling. After that, the terms are clustered with `hclust()` and the dendrogram is cut into 10 clusters. The agglomeration method is set to `ward`, which denotes the increase in variance when two clusters are merged. Some other options are single linkage, complete linkage, average linkage, median and centroid. Details about different agglomeration methods can be found in data mining text books (Han and Kamber, 2000; Hand et al., 2001; Witten and Frank, 2005).

```
> # remove sparse terms

> myTdm2 <- removeSparseTerms(myTdm, sparse=0.95)

> m2 <- as.matrix(myTdm2)

> # cluster terms

> distMatrix <- dist(scale(m2))

> fit <- hclust(distMatrix, method="ward")

> plot(fit)

> # cut tree into 10 clusters

> rect.hclust(fit, k=10)
```

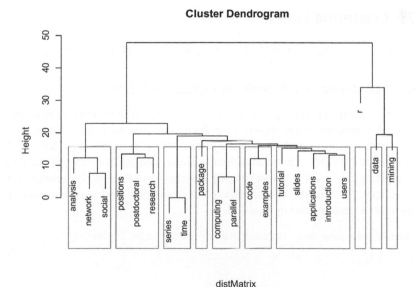

Figure 10.3 Clustering of words.

```
> (groups <- cutree(fit, k=10))
```

analysis	applications	code	computing	data	examples
1	2	3	4	5	3
introduction	mining	network	package	parallel	positions
2	6	1	7	4	8
postdoctoral	r	research	series	slides	social
8	9	8	10	2	1
time	tutorial	users			
10	2	2			

In the above dendrogram (see Figure 10.3), we can see the topics in the tweets. Words "analysis," "network" and "social" are clustered into one group, because there are a couple of tweets on social network analysis. The second cluster from left comprises "positions," "postdoctoral" and "research," and they are clustered into one group because of tweets on vacancies of research and postdoctoral positions. We can also see cluster on time series, R packages, parallel computing, R codes and examples, and tutorial and slides. The rightmost three clusters consists of "r," "data" and "mining," which are the keywords of @RDataMining tweets.

10.8 Clustering Tweets

Tweets are clustered below with the k-means and the k-medoids algorithms.

10.8.1 Clustering Tweets with the k-Means Algorithm

We first try k-means clustering, which takes the values in the matrix as numeric. We transpose the term-document matrix to a document-term one. The tweets are then clustered with kmeans() with the number of clusters set to eight. After that, we check the popular words in every cluster and also the cluster centers. Note that a fixed random seed is set with set.seed() before running kmeans(), so that the clustering result can be reproduced. It is for the convenience of book writing, and it is unnecessary for readers to set a random seed in their code.

```
> # transpose the matrix to cluster documents (tweets)

> m3 <- t(m2)

> # set a fixed random seed

> set.seed(122)

> # k-means clustering of tweets

> k <- 8
```

```
> kmeansResult <- kmeans(m3, k)

> # cluster centers

> round(kmeansResult$centers, digits=3)
```

	analysis	applications	code	computing	data	examples	introduction	mining
1	0.040	0.040	0.240	0.000	0.040	0.320	0.040	0.120
2	0.000	0.158	0.053	0.053	1.526	0.105	0.053	1.158
3	0.857	0.000	0.000	0.000	0.000	0.071	0.143	0.071
4	0.000	0.000	0.000	1.000	0.000	0.000	0.000	0.000
5	0.037	0.074	0.019	0.019	0.426	0.037	0.093	0.407
6	0.000	0.000	0.000	0.000	0.000	0.100	0.000	0.000
7	0.533	0.000	0.067	0.000	0.333	0.200	0.067	0.200
8	0.000	0.111	0.000	0.000	0.556	0.000	0.000	0.111

	network	package	parallel	positions	postdoctoral	r	research	series	slides
1	0.080	0.080	0.000	0.000	0.000	1.320	0.000	0.040	0.000
2	0.000	0.368	0.053	0.000	0.000	0.947	0.053	0.000	0.053
3	1.000	0.071	0.000	0.143	0.143	0.214	0.071	0.000	0.071
4	0.000	0.125	0.750	0.000	0.000	1.000	0.000	0.000	0.125
5	0.000	0.000	0.000	0.093	0.093	0.000	0.000	0.019	0.074
6	0.000	1.200	0.100	0.000	0.000	0.600	0.100	0.000	0.100
7	0.067	0.000	0.000	0.000	0.000	1.000	0.000	0.400	0.533
8	0.000	0.000	0.000	0.444	0.444	0.000	1.333	0.000	0.000

	social	time	tutorial	users
1	0.000	0.040	0.200	0.160
2	0.000	0.000	0.000	0.158
3	0.786	0.000	0.286	0.071
4	0.000	0.000	0.125	0.250
5	0.000	0.019	0.111	0.019
6	0.000	0.000	0.100	0.100
7	0.000	0.400	0.000	0.400
8	0.111	0.000	0.000	0.000

To make it easy to find what the clusters are about, we then check the top three words in every cluster.

```
> for (i in 1:k) {

+ cat(paste("cluster ", i, ": ", sep=""))

+ s <- sort(kmeansResult$centers[i,], decreasing=T)

+ cat(names(s)[1:3], "\ n")

+ # print the tweets of every cluster

+ # print(rdmTweets[which(kmeansResult$cluster==i)])

+ }
 cluster 1:   r examples code
 cluster 2:   data mining r
 cluster 3:   network analysis social
 cluster 4:   computing r parallel
 cluster 5:   data mining tutorial
 cluster 6:   package r examples
 cluster 7:   r analysis slides
 cluster 8:   research data positions
```

From the above top words and centers of clusters, we can see that the clusters are of different topics. For instance, cluster 1 focuses on R codes and examples, cluster 2 on data mining with R, cluster 4 on parallel computing in R, cluster 6 on R packages and cluster 7 on slides of time series analysis with R. We can also see that, all clusters, except for cluster 3, 5, & 8, focus on R. Cluster 3, 5, & 8 are about general information on data mining and are not limited to R. Cluster 3 is on social network analysis, cluster 5 on data mining tutorials, and cluster 8 on positions for data mining research.

10.8.2 Clustering Tweets with the k-Medoids Algorithm

We then try k-medoids clustering with the Partitioning Around Medoids (PAM) algorithm, which uses medoids (representative objects) instead of means to represent clusters. It is more robust to noise and outliers than k-means clustering, and provides a display of the silhouette plot to show the quality of clustering. In the example below, we use function pamk() from package *fpc* (Hennig, 2010), which calls the function pam() with the number of clusters estimated by optimum average silhouette.

```
> library(fpc)

> # partitioning around medoids with estimation of number of
  clusters
```

```
> pamResult <- pamk(m3, metric="manhattan")

> # number of clusters identified

> (k <- pamResult$nc)

[1] 9

> pamResult <- pamResult$pamobject

> # print cluster medoids

> for (i in 1:k) {

+    cat(paste("cluster", i, ": "))

+    cat(colnames(pamResult$medoids)
      [which(pamResult$medoids[i,]==1)], "\n")

+    # print tweets in cluster i

+    # print(rdmTweets[pamResult$clustering==i])

+ }

 cluster 1:   data positions research
 cluster 2:   computing parallel r
 cluster 3:   mining package r
 cluster 4:   data mining
 cluster 5:   analysis network social tutorial
 cluster 6:   r
 cluster 7:
 cluster 8:   examples r
 cluster 9:   analysis mining series time users


> # plot clustering result

> layout(matrix(c(1,2),2,1)) # set to two graphs per page

> plot(pamResult, color=F, labels=4, lines=0, cex=.8, col.clus=1,

+     col.p=pamResult$clustering)

> layout(matrix(1)) # change back to one graph per page
```

clusplot(pam(x = sdata, k = k, metric = "manhattan"))

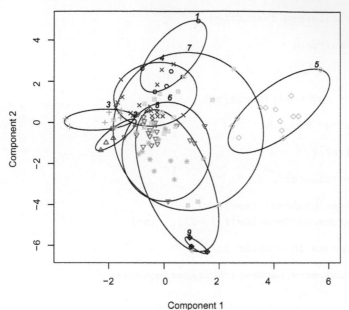

Component 1

These two components explain 24.81 % of the point variability.

Silhouette plot of pam(x = sdata, k = k, metric = "manhattan")

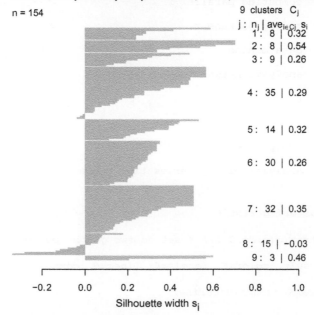

Average silhouette width : 0.29

Figure 10.4 Clusters of tweets.

In Figure 10.4, the first chart is a 2D "clusplot" (clustering plot) of the k clusters, and the second one shows their silhouettes. With the silhouette, a large s_i (almost 1) suggests that the corresponding observations are very well clustered, a small s_i (around 0) means that the observation lies between two clusters, and observations with a negative s_i are probably placed in the wrong cluster. The average silhouette width is 0.29, which suggests that the clusters are not well separated from one another.

The above results and Figure 10.4 show that there are nine clusters of tweets. Clusters 1, 2, 3, 5, and 9 are well separated groups, with each of them focusing on a specific topic. Cluster 7 is composed of tweets not fitted well into other clusters, and it overlaps all other clusters. There is also a big overlap between cluster 6 and 8, which is understandable from their medoids. Some observations in cluster 8 are of negative silhouette width, which means that they may fit better in other clusters than cluster 8.

To improve the clustering quality, we have also tried to set the range of cluster numbers `krange=2:8` when calling `pamk()`, and in the new clustering result, there are eight clusters, with the observations in the above cluster 8 assigned to other clusters, mostly to cluster 6. The results are not shown in this book, and readers can try it with the code below.

```
> pamResult2 <- pamk(m3, krange=2:8, metric="manhattan")
```

10.9   Packages, Further Readings, and Discussions

In addition to frequent terms, associations and clustering demonstrated in this chapter, some other possible analysis on the above Twitter text is graph mining and social network analysis. For example, a graph of words can be derived from a document-term matrix, and then we can use techniques for graph mining to find links between words and groups of words. A graph of tweets (documents) can also be generated and analyzed in a similar way. It can also be presented and analyzed as a bipartite graph with two disjoint sets of vertices, that is, words and tweets. We will demonstrate social network analysis on the Twitter data in Chapter 11 Social Network Analysis.

Some R packages for text mining are listed below.

- Package *tm* (Feinerer, 2012): A framework for text mining applications within R.

- Package *tm.plugin.mail* (Feinerer, 2010): Text Mining E-Mail Plug-In. A plug-in for the tm text mining framework providing mail handling functionality.

- package *textcat* (Hornik et al., 2012) provides n-Gram Based Text Categorization.

- *lda* (Chang, 2011) fits topic models with LDA (latent Dirichlet allocation)

- *topicmodels* (Grün and Hornik, 2011) fits topic models with LDA and CTM (correlated topics model)

For more information and examples on text mining with R, some online resources are:

- *Introduction to the tm Package – Text Mining in R*
  http://cran.r-project.org/web/packages/tm/vignettes/tm.pdf

- *Text Mining Infrastructure in R* (Feinerer, 2008)
  http://www.jstatsoft.org/v25/i05

- Text Mining Handbook
  http://www.casact.org/pubs/forum/10spforum/Francis_Flynn.pdf

- Distributed Text Mining in R http://epub.wu.ac.at/3034/

- Text mining with Twitter and R
  http://heuristically.wordpress.com/2011/04/08/
  text-data-mining-twitter-r/

11 Social Network Analysis

This chapter presents examples of social network analysis with R, specifically, with package *igraph* (Csardi and Nepusz, 2006). The data to analyze is Twitter text data used in Chapter 10 Text Mining. Putting it in a general scenario of social networks, the terms can be taken as people and the tweets as groups on LinkedIn,[1] and the term-document matrix can then be taken as the group membership of people.

In this chapter, we first build a network of terms based on their co-occurrence in the same tweets, and then build a network of tweets based on the terms shared by them. At last, we build a two-mode network composed of both terms and tweets. We also demonstrate some tricks to plot nice network graphs. Some codes in this chapter are based on the examples at `http://www.stanford.edu/~messing/Affiliation%20Data.html`.

11.1 Network of Terms

In this section, we will build a network of terms based on their co-occurrence in tweets. At first, a term-document matrix, `termDocMatrix`, is loaded into R, which is actually a copy of `m2`, an R object from Chapter 10 Text Mining (see page 115). After that, it is transformed into a term-term adjacency matrix, based on which a graph is built. Then we plot the graph to show the relationship between frequent terms, and also make the graph more readable by setting colors, font sizes, and transparency of vertices and edges.

```
> # load termDocMatrix
> load("./data/termDocMatrix.rdata")
> # inspect part of the matrix
> termDocMatrix[5:10,1:20]
```

	Docs																			
Terms	1	2	3	4	5	6	7	8	9	10	11	12	13	14	15	16	17	18	19	20
data	1	1	0	0	2	0	0	0	0	0	1	2	1	1	1	0	1	0	0	0
Examples	0	0	0	0	0	0	0	0	0	0	0	0	0	0	0	0	0	0	0	0
introduction	0	0	0	0	0	0	0	0	0	0	0	0	0	0	0	0	0	0	0	1
mining	0	0	0	0	0	0	0	0	0	0	0	1	1	0	1	0	0	0	0	0
network	0	0	0	0	0	0	0	0	0	0	0	0	0	0	0	1	0	1	1	1
package	0	0	0	1	1	0	0	0	0	0	0	1	0	0	0	0	0	0	0	0

[1] `http://www.linkedin.com`

R and Data Mining. DOI: http://dx.doi.org/10.1016/B978-0-12-396963-7.00011-8

```
> # change it to a Boolean matrix
> termDocMatrix[termDocMatrix>=1] <- 1
> # transform into a term-term adjacency matrix
> termMatrix <- termDocMatrix %*% t(termDocMatrix)
> # inspect terms numbered 5 to 10
> termMatrix[5:10,5:10]
```

	Terms					
Terms	data	examples	introduction	mining	network	package
data	53	5	2	34	0	7
Examples	5	17	2	5	2	2
introduction	2	2	10	2	2	0
mining	34	5	2	47	1	5
network	0	2	2	1	17	1
package	7	2	0	5	1	21

In the above code, `%*%` is an operator for the product of two matrices, and `t()` transposes a matrix. Now we have built a term-term adjacency matrix, where the rows and columns represent terms, and every entry is the number of concurrences of two terms. Next we can build a graph with `graph.adjacency()` from package *igraph*.

```
> library(igraph)

> # build a graph from the above matrix

> g <- graph.adjacency(termMatrix, weighted=T, mode="undirected")

> # remove loops

> g <- simplify(g)

> # set labels and degrees of vertices

> V(g)$label <- V(g)$name

> V(g)$degree <- degree(g)
```

After that, we plot the network with `layout.fruchterman.reingold` (see Figure 11.1).

```
> # set seed to make the layout reproducible

> set.seed(3952)

> layout1 <- layout.fruchterman.reingold(g)

> plot(g, layout=layout1)
```

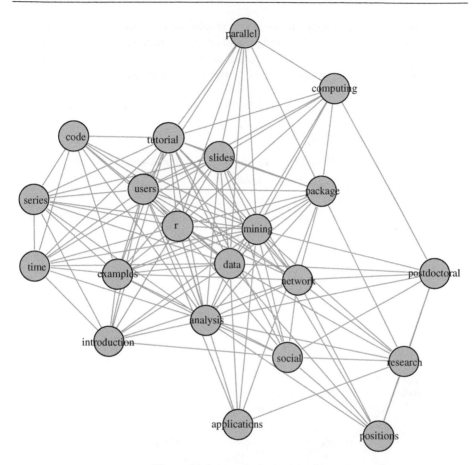

Figure 11.1 A network of terms—I.

In the above code, the layout is kept as layout1, so that we can plot the graph in the same layout later.

A different layout can be generated with the first line of code below. The second line produces an interactive plot, which allows us to manually rearrange the layout. Details about other layout options can be obtained by running ?igraph::layout in R.

```
> plot(g, layout=layout.kamada.kawai)
> tkplot(g, layout=layout.kamada.kawai)
```

We can also save the network graph into a.PDF file with the code below.

```
> pdf("term-network.pdf")
> plot(g, layout=layout.fruchterman.reingold)
> dev.off()
```

Next, we set the label size of vertices based on their degrees, to make important terms stand out. Similarly, we also set the width and transparency of edges based on their weights. This is useful in applications where graphs are crowded with many vertices and edges. In the code below, the vertices and edges are accessed with `V()` and `E()`. Function `rgb(red, green, blue, alpha)` defines a color, with an `alpha` transparency. With the same layout as Figure 11.1, we plot the graph again (see Figure 11.2).

```
> V(g)$label.cex <- 2.2 * V(g)$degree / max(V(g)$degree)+ .2
> V(g)$label.color <- rgb(0, 0,.2,.8)
> V(g)$frame.color <- NA
> egam <- (log(E(g)$weight)+.4) / max(log(E(g)$weight)+.4)
> E(g)$color <- rgb(.5,.5, 0, egam)
> E(g)$width <- egam
> # plot the graph in layout1
> plot(g, layout=layout1)
```

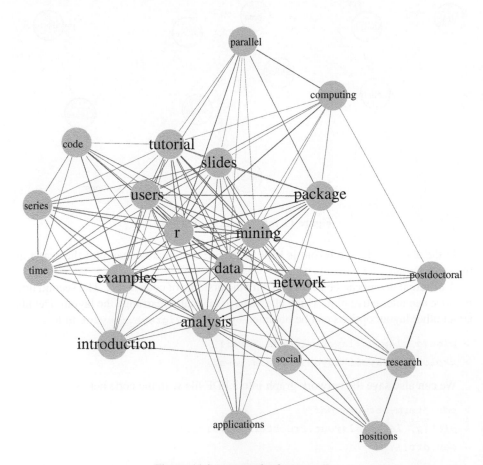

Figure 11.2 A network of terms—II.

11.2   Network of Tweets

Similar to the previous section, we can also build a graph of tweets based on the number of terms that they have in common. Because most tweets contain one or more words from "r", "data", and "mining", most tweets are connected with others and the graph of tweets is very crowded. To simplify the graph and find relationship between tweets beyond the above three keywords, we remove the three words before building a graph.

```
> # remove "r", "data" and "mining"

> idx <- which(dimnames(termDocMatrix)$Terms %in% c("r", "data",
  "mining"))

> M <- termDocMatrix[-idx,]

> # build a tweet-tweet adjacency matrix

> tweetMatrix <- t(M) %*% M

> library(igraph)

> g <- graph.adjacency(tweetMatrix, weighted=T,
  mode="undirected")

> V(g)$degree <- degree(g)

> g <- simplify(g)

> # set labels of vertices to tweet IDs

> V(g)$label <- V(g)$name

> V(g)$label.cex <- 1

> V(g)$label.color <- rgb(.4, 0, 0,.7)

> V(g)$size <- 2

> V(g)$frame.color <- NA
```

Next, we have a look at the distribution of degree of vertices and the result is shown in Figure 11.3. We can see that there are around 40 isolated vertices (with a degree of zero). Note that most of them are caused by the removal of the three keywords, "r", "data", and "mining".

```
> barplot(table(V(g)$degree))
```

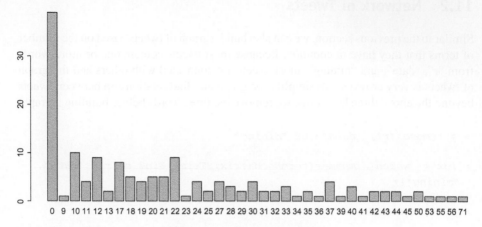

Figure 11.3 Distribution of degree.

With the code below, we set vertex colors based on degree, and set labels of isolated vertices to tweet IDs and the first 20 characters of every tweet. The labels of other vertices are set to tweet IDs only, so that the graph will not be overcrowded with labels. We also set the color and width of edges based on their weights. The produced graph is shown in Figure 11.4.

```
> idx <- V(g)$degree == 0

> V(g)$label.color[idx] <- rgb(0, 0, .3, .7)

> # load twitter text

> library(twitteR)

> load(file = "data/rdmTweets.RData")

> # convert tweets to a data frame

> df <- do.call("rbind", lapply(rdmTweets, as.data.frame))

> # set labels to the IDs and the first 20 characters of tweets

> V(g)$label[idx] <- paste(V(g)$name[idx], substr(df$text[idx],
  1, 20), sep=": ")

> egam <- (log(E(g)$weight)+.2) / max(log(E(g)$weight)+.2)

> E(g)$color <- rgb(.5, .5, 0, egam)
```

```
> E(g)$width <- egam

> set.seed(3152)

> layout2 <- layout.fruchterman.reingold(g)

> plot(g, layout=layout2)
```

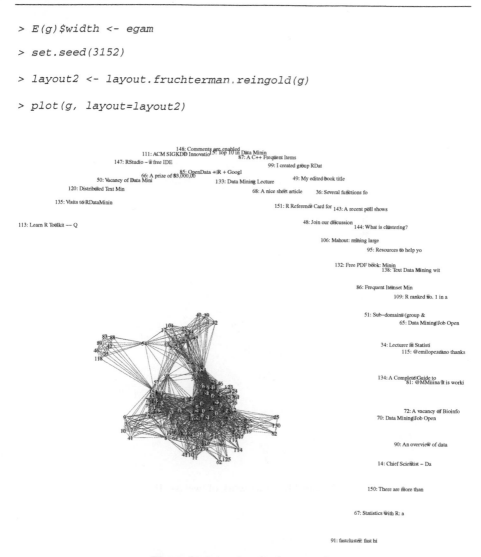

Figure 11.4 A network of tweets—I.

The vertices in crescent are isolated from all others, and next we remove them from graph with function delete.vertices() and re-plot the graph (see Figure 11.5).

```
> g2 <- delete.vertices(g,V(g)[degree(g)==0])

> plot(g2, layout=layout.fruchterman.reingold)
```

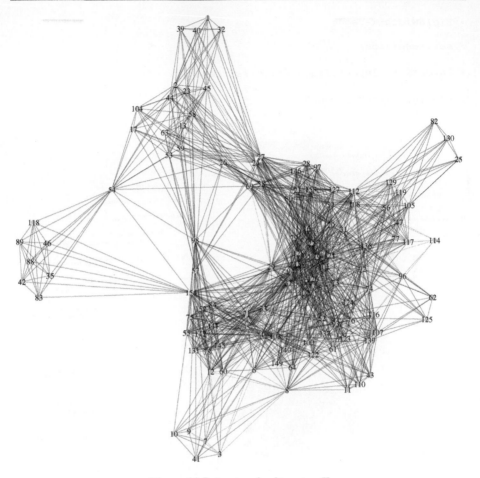

Figure 11.5 A network of tweets—II.

Similarly, we can also remove edges with low degrees to simplify the graph. Below with function delete.edges(), we remove edges which have weight of one. After removing edges, some vertices become isolated and are also removed. The produced graph is shown in Figure 11.6.

```
> g3 <- delete.edges(g, E(g)[E(g)$weight <= 1])

> g3 <- delete.vertices(g3, V(g3)[degree(g3) == 0])

> plot(g3, layout=layout.fruchterman.reingold)
```

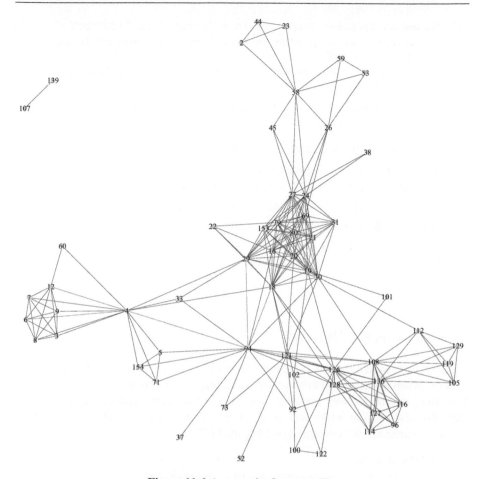

Figure 11.6 A network of tweets—III.

In Figure 11.6, there are some groups (or cliques) of tweets. Let's have a look at the group in the middle left of the figure.

```
> df$text[c(7,12,6,9,8,3,4)]
```

```
[7] State of the Art in Parallel Computing with R
    http://t.co/zmClglqi
[12] The R Reference Card for Data Mining is updated with
    functions & packages for handling big data & parallel computing.
    http://t.co/FHoVZCyk
[6] Parallel Computing with R using snow and snowfall
    http://t.co/nxp8EZpv
[9] R with High Performance Computing: Parallel processing and
    large memory  http://t.co/XZ3ZZBRF
```

[8] Slides on Parallel Computing in R http://t.co/AdDVxbOY
[3] Easier Parallel Computing in R with snowfall and sfCluster
  http://t.co/BPcinvzK
[4] Tutorial: Parallel computing using R package snowfall
  http://t.co/CHBCyr76

   We can see that tweets 7, 12, 6, 9, 8, 3, 4 are on parallel computing with R. We can
also see some other groups below:

- Tweets 4, 33, 94, 29, 18, and 92: tutorials for R;
- Tweets 4, 5, 154, and 71: R packages;
- Tweets 126, 128, 108, 136, 127, 116, 114, and 96: time series analysis;
- Tweets 112, 129, 119, 105, 108, and 136: R code examples; and
- Tweets 27, 24, 22,153, 79, 69, 31, 80, 21, 29, 16, 20, 18, 19, and 30: social network
  analysis.

Tweet 4 lies between multiple groups, because it contains keywords "parallel computing", "tutorial", and "package".

11.3   Two-Mode Network

In this section, we will build a two-mode network, which is composed of two types of
vertices: tweets and terms. At first, we generate a graph g directly from termDocMatrix.
After that, different colors and sizes are assigned to term vertices and tweet vertices. We also set the width and color of edges. The graph is then plotted with
layout.fruchterman.reingold (see Figure 11.7).

```
> # create a graph
> g <- graph.incidence(termDocMatrix, mode=c("all"))
> # get index for term vertices and tweet vertices
> nTerms <- nrow(M)
> nDocs <- ncol(M)
> idx.terms <- 1:nTerms
> idx.docs <- (nTerms+1):(nTerms+nDocs)
> # set colors and sizes for vertices
> V(g)$degree <- degree(g)
> V(g)$color[idx.terms] <- rgb(0, 1, 0,.5)
> V(g)$size[idx.terms] <- 6
> V(g)$color[idx.docs] <- rgb(1, 0, 0,.4)
> V(g)$size[idx.docs] <- 4
> V(g)$frame.color <- NA
> # set vertex labels and their colors and sizes
> V(g)$label <- V(g)$name
```

```
> V(g)$label.color <- rgb(0, 0, 0, 0.5)
> V(g)$label.cex <- 1.4*V(g)$degree/max(V(g)$degree) + 1
> # set edge width and color
> E(g)$width <-.3
> E(g)$color <- rgb(.5, .5, 0, .3)
> set.seed(958)
> plot(g, layout=layout.fruchterman.reingold)
```

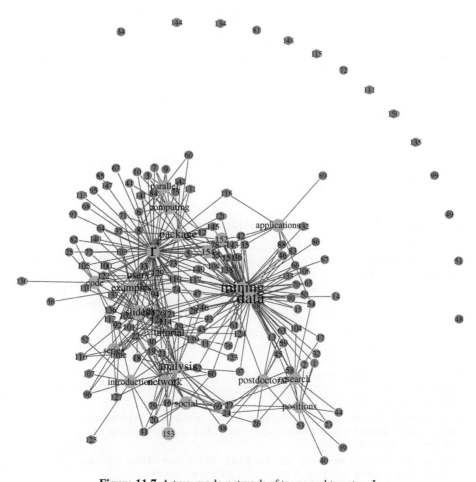

Figure 11.7 A two-mode network of terms and tweets—I.

Figure 11.7 shows that most tweets are around two centers, "r" and "data mining". Next, let's have a look at which tweets are about "r". In the code below, `nei("r")` returns all vertices which are neighbors of vertex "r".

```
> V(g)[nei("r")]
```

```
Vertex sequence:


 [1] "3"   "4"   "5"   "6"   "7"   "8"   "9"   "10"  "12"  "19"  "21"  "22"
[13] "25"  "28"  "30"  "33"  "35"  "36"  "41"  "42"  "55"  "64"  "67"  "68"
[25] "73"  "74"  "75"  "77"  "78"  "82"  "84"  "85"  "91"  "92"  "94"  "95"
[37] "100" "101" "102" "105" "108" "109" "110" "112" "113" "114" "117" "118"
[49] "119" "120" "121" "122" "126" "128" "129" "131" "136" "137" "138" "140"
[61] "141" "142" "143" "145" "146" "147" "149" "151" "152" "154"
```

An alternative way is using function neighborhood() as below.

```
> V(g)[neighborhood(g, order=1, "r")[[1]]]
```

We can also have a further look at which tweets contain all three terms: "r", "data", and "mining".

```
> (rdmVertices <- V(g)[nei("r") & nei("data") & nei("mining")])

  Vertex sequence:


 [1] "12"  "35"  "36" "42" "55" "78" "117" "119" "138" "143" "149" "151"
[13] "152" "154"
```

```
> df$text[as.numeric(rdmVertices$label)]
```

[12] The R Reference Card for Data Mining is updated with
   functions & packages for handling big data & parallel computing.
   http://t.co/FHoVZCyk

[35] Call for reviewers: Data Mining Applications with R. Pls
   contact me if you have experience on the topic. See details at
   http://t.co/rcYIXfnp

[36] Several functions for evaluating performance of
   classification models added to R Reference Card for Data Mining:
   http://t.co/FHoVZCyk

[42] Call for chapters: Data Mining Applications with R, an
   edited book to be published by Elsevier. Proposal due 30 April.
   http://t.co/HPaBSbRa

[55] Some R functions and packages for outlier detection
   have been added to R Reference Card for Data Mining at
   http://t.co/FHoVZCyk.

[78] Access large amounts of Twitter data for data mining
   and other tasks within R via the twitteR package.
   http://t.co/ApbAbnxs

[117] My document, R and Data Mining—Examples and Case Studies,
   is scheduled to be published by Elsevier in mid 2012.
   http://t.co/BcqwQ1n

[119] Lecture Notes on data mining course at CMU, some of which
   contain R code examples. http://t.co/7YY73OW

[138] Text Data Mining with Twitter and R. http://t.co/a50ySNq

```
[143] A recent poll shows that R is the 2nd popular tool used
    for data mining. See Poll: Data Mining/Analytic Tools Used
    http://t.co/ghpbQXq
```

To make it short, only the first 10 tweets are displayed in the above result. In the above code, df is a data frame which keeps tweets of RDataMining, and details of it can be found in Section 10.2.

Next, we remove "r", "data", and "mining" to show the relationship between tweets with other words. Isolated vertices are also deleted from graph.

```
> idx <- which(V(g)$name %in% c("r", "data", "mining"))
> g2 <- delete.vertices(g, V(g)[idx-1])
> g2 <- delete.vertices(g2, V(g2)[degree(g2)==0])
> set.seed(209)
> plot(g2, layout=layout.fruchterman.reingold)
```

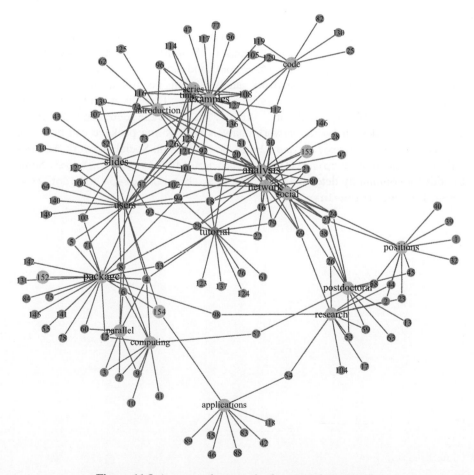

Figure 11.8 A two-mode network of terms and tweets—II.

From Figure 11.8, we can clearly see groups of tweets and their keywords, such as time series, social network analysis, parallel computing, and postdoctoral and research positions, which are similar to the result presented at the end of Section 11.2.

11.4 Discussions and Further Readings

In this chapter, we have demonstrated how to find groups of tweets and some topics in the tweets with package *igraph*. Similar analysis can also be achieved with package *sna* (Butts, 2010). There are also packages designed for topic modeling, such as packages *lda* (Chang, 2011) and *topicmodels* (Grün and Hornik, 2011).

For readers interested in social network analysis with R, there are some further readings. Some examples on social network analysis with the *igraph* package (Csardi and Nepusz, 2006) are available as tutorial on *Network Analysis with Package igraph* at `http://igraph.sourceforge.net/igraphbook/` and R for Social Network Analysis at `http://www.stanford.edu/~messing/RforSNA.html`. There is a detailed introduction to Social Network Analysis with package *sna* (Butts, 2010) at `http://www.jstatsoft.org/v24/i06/paper`. A *statnet* Tutorial is available at `http://www.jstatsoft.org/v24/i09/paper` and more resources on using *statnet* (Handcock et al., 2003) for network analysis can be found at `http://csde.washington.edu/statnet/resources.shtml`. There is a short tutorial on package *network* (Butts, 2012) at `http://sites.stat.psu.edu/~dhunter/Rnetworks/`. Slides on Social network analysis with R *sna* package can be found at `http://user2010.org/slides/Zhang.pdf`. slides on Social Network Analysis in R can be found at `http://files.meetup.com/1406240/sna_in_R.pdf`. Some R codes for community detection are available at `http://igraph.wikidot.com/community-detection-in-r`.

12 Case Study I: Analysis and Forecasting of House Price Indices

This chapter presents a case study on analyzing and forecasting of House Price Indices (HPI). It demonstrates data import from a CSV file, descriptive analysis of HPI time series data, and decomposition and forecasting of the data. The data used in this study are Canberra house price trading indices from Residex.[1]

Note that this study is to demonstrate how to use R to study time series data for research purpose only. The analysis of property market may involve many other factors not mentioned in this chapter, such as economic environment, population size, CPI (Consumer Price Index), and government policy, and the readers should make their own judgment if interested in property investment.

12.1 Importing HPI Data

The data records the HPIs at the end of every month from January 1990 to January 2011, and the first four lines of data are shown below as examples:

```
31-Jan-90,1.00763

28-Feb-90,1.01469

31-Mar-90,1.02241

30-Apr-90,1.03062
```

At first, the data are read from a CSV file with `read.csv()`, and then names are assigned to columns of data frame `houseIndex`. After that, the dates are converted with function `strptime()` from character to "POSIXlt" to extract year and month.

```
> # import data

> filepath <- "./data/"

> filename <- "House-index-canberra.csv"
```

[1] http://www.residex.com.au

R and Data Mining. DOI: http://dx.doi.org/10.1016/B978-0-12-396963-7.00012-X

```
> houseIndex <- read.csv(paste(filepath, filename, sep=""),
  header=FALSE)

> names(houseIndex) <- c("date", "index")

> n <- nrow(houseIndex)

> # check start date and end date

> cat(paste("HPI from", houseIndex$date[1], "to",
  houseIndex$date[n], "\n"))

HPI from 31-Jan-90 to 31-Jan-11

> # extract year and month

> dates <- strptime(houseIndex$date, format="%d-%b-%y")

> houseIndex$year <- dates$year + 1900

> houseIndex$month <- dates$mon + 1

> fromYear <- houseIndex$year[1]
```

An alternative way for the above format conversion is to use function as.Date()
as follows:

```
> dates <- as.Date(houseIndex$date, format="%d-%b-%y")

> houseIndex$year <- as.numeric(format(dates, "%y"))

> houseIndex$month <- as.numeric(format(dates, "%m"))
```

12.2   Exploration of HPI Data

The data are explored with various plots of HPI and its variations over the years. First,
a chart is drawn below to show the changes of HPI from 1990 to 2011 (see Figure 12.1).

```
> plot(houseIndex$index, pty=1, type="l", lty="solid", xaxt="n",
  xlab="", ylab="Index",

+     main=paste("HPI(Canberra) - Since ", fromYear, sep=""))

> # draw tick-marks at 31 Jan of every year

> nYear <- ceiling(n/12)

> posEveryYear <- 12 * (1:nYear) - 11

> axis(1, labels=houseIndex$date[posEveryYear], las=3,
  at=posEveryYear)
```

```
> # add horizontal reference lines

> abline(h=1:4, col="gray", lty="dotted")

> # draw a vertical reference line every five years

> posEvery5years <- 12 * (5* 1:ceiling(nYear/5) - 4) - 11

> abline(v=posEvery5years, col="gray", lty="dotted")
```

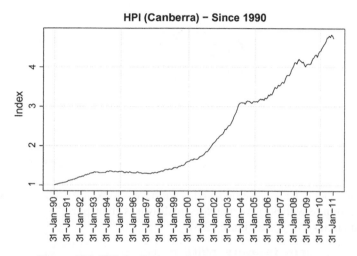

Figure 12.1 HPIs in Canberra from Jan. 1990 to Jan. 2011.

Although a reference grid can be added with grid(), the positions of the lines generated with it are not necessarily aligned with the beginning of every year. Therefore, in the above example, the positions of reference lines are calculated first and then drawn with abline().

Let us have a look at the increase of HPI in every month, which is calculated as delta (see Figure 12.2).

```
> houseIndex$delta <- houseIndex$index - c(1, houseIndex$index[-n])

> plot(houseIndex$delta, main="Increase in HPI", xaxt="n",
  xlab="")

> axis(1, labels=houseIndex$date[posEveryYear], las=3,
  at=posEveryYear)

> # add a reference line

> abline(h=0, lty="dotted")
```

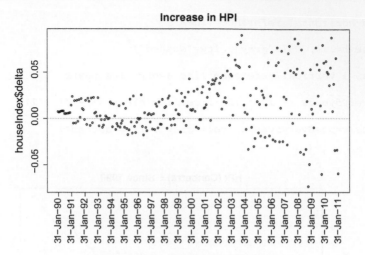

Figure 12.2 Monthly increase of HPI.

It seems from Figure 12.2 that HPI fluctuated more after 2003 than before. However, it may be simply because of the increase of HPI from one in 1990 to around five in 2011.

To further check the fluctuation in HPI, we have a look at its ratio of increase per month. The months with positive increase are drawn as green plus ("+"), while negative ones as red circles ("o") (see Figure 12.3).

```
> # increase ratio in every month

> houseIndex$rate <- houseIndex$index/c(1, houseIndex$index[-n]) - 1

> # percentage of months having positive increases in HPI

> 100 * sum(houseIndex$rate>0)/n

[1] 67.58893

> # use ifelse() to set positive values to green and and negative
  ones to red

> plot(houseIndex$rate, xaxt="n", xlab="", ylab="HPI Increase
  Rate", col=ifelse(houseIndex$rate>0,"green","red"),

+     pch=ifelse(houseIndex$rate>0,"+","o"))

> axis(1, labels=houseIndex$date[posEveryYear], las=3,
  at=posEveryYear)
```

```
> abline(h=0, lty="dotted")
```

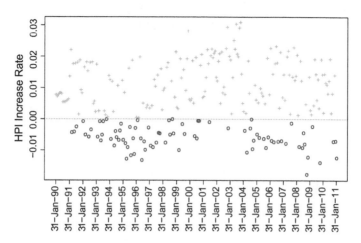

Figure 12.3 Monthly increase rate of HPI.

From Figure 12.3, we can see that: (1) there are more increases than decreases; (2) the increase rates (shown as "+" in green) are generally bigger than decrease rates ("o" in red); most increase rates are between 0 and 2%, and most decrease rates are between 0 and 1%; and (3) there are two periods with big decreases: 1995–1996 and 2008–2009; meanwhile, a period having biggest increases is 2002–2003.

Alternatively, we make a table of increase rate, with each row standing for a month and each column for a year, and then show the monthly increase rates with a grouped bar chart (see Figure 12.4). In the bar chart, the columns are portrayed as juxtaposed bars instead of stacked bars by setting "beside=TRUE", and the space between groups are set with "space=c(0,2)".

```
> rateMatrix <- xtabs(rate ~ month + year, data=houseIndex)

> # show the first four years, rounded to 4 decimal places

> round(rateMatrix[,1:4], digits=4)
```

	year			
month	1990	1991	1992	1993
1	0.0076	0.0134	−0.0007	0.0172
2	0.0070	0.0219	0.0164	−0.0057
3	0.0076	−0.0043	−0.0050	0.0023
4	0.0080	0.0174	0.0180	−0.0007
5	0.0082	−0.0041	0.0151	−0.0069
6	0.0079	0.0176	−0.0057	−0.0051

```
 7     0.0052    -0.0025     0.0178    -0.0008
 8     0.0053     0.0179    -0.0034     0.0023
 9     0.0053     0.0013     0.0180     0.0010
10     0.0055     0.0186     0.0046    -0.0001
11     0.0058     0.0091     0.0055     0.0004
12     0.0061     0.0081     0.0021     0.0136
```

```
> # plot a grouped barchart:

> barplot(rateMatrix, beside=TRUE, space=c(0,2),

+          col=ifelse(rateMatrix>0,"lightgreen","lightpink"),

+          ylab="HPI Increase Rate", cex.names=1.2)
```

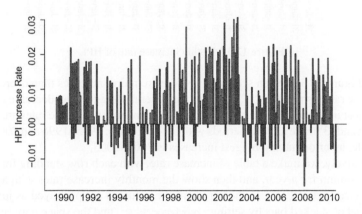

Figure 12.4 A bar chart of monthly HPI increase rate.

Figures 12.5, 12.6 and 12.7 show respectively the number of months with HPI increases over the years and the yearly/monthly average increase rates. Functions colSums(), colMeans(), and rowMeans() are used to calculate the corresponding values based on rateMatrix.

```
> numPositiveMonths <- colSums(rateMatrix > 0)

> barplot(numPositiveMonths, xlab="Year", ylab="Number of Months
  with Increased HPI")

> yearlyMean <- colMeans(rateMatrix)

> barplot(yearlyMean, main="Yearly Average Increase Rates of
  HPI", col=ifelse(yearlyMean>0,"lightgreen","lightpink"),
  xlab="Year")

> monthlyMean <- rowMeans(rateMatrix)
```

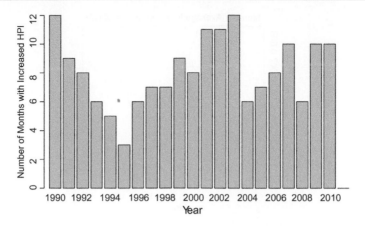

Figure 12.5 Number of months with increased HPI.

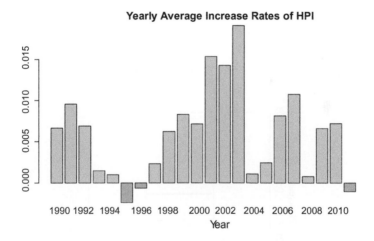

Figure 12.6 Yearly average increase rates of HPI.

```
> plot(names(monthlyMean), monthlyMean, type="b", xlab="Month",

+       main="Monthly Average Increase Rates of HPI")
```

Next, the distribution of increase rate is checked. Function `summary()` returns the minimum, maximum, mean, median, and the first (25%) and third quartiles (75%) of data. A box-and-whisker plot (see Figure 12.8), generated by `boxplot()`, shows the median, and the first and third quartiles. The bar in the middle is the median. The box shows the interquartile range (IQR) , which is the range between the first and third quartiles.

```
> summary(houseIndex$rate)
```

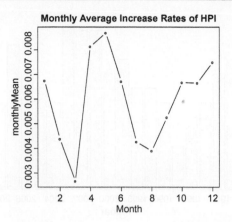

Figure 12.7 Monthly average increase rates of HPI.

Min.	1st Qu.	Median	Mean	3rd Qu.	Max.
-0.017710	-0.002896	0.005840	0.006222	0.014210	0.030700

```
> boxplot(houseIndex$rate, ylab="HPT Increase Rate"
```

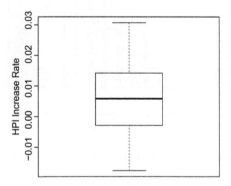

Figure 12.8 Distribution of HPI increase rate.

We then further check the distribution of increase rate for every year (see Figure 12.9) and also for every month (see Figure 12.10) with grouped boxplots.

```
> boxplot(rate ~ year, data=houseIndex, xlab="year", ylab="HPI
  Increase Rate")
```

```
> boxplot(rate ~ month, data=houseIndex, xlab="month", ylab="HPI
  Increase Rate")
```

Figure 12.10 shows that April and May are the months when house prices increase fastest, because in the two months, the median increase rates are high and most increase

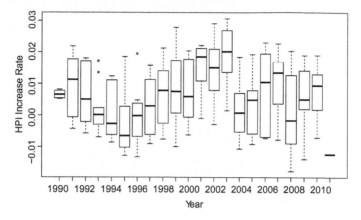

Figure 12.9 Distribution of HPI increase rate per year.

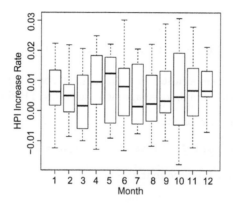

Figure 12.10 Distribution of HPI increase rate per month.

rates are positive. Some other months with high increase rates are June, January, and December. The increases of house price are low in March, July, and August.

12.3 Trend and Seasonal Components of HPI

After the above data exploration and basic analysis, we decompose the data to find trend and seasonal factors, and then make a forecast of HPI. For more details about time series decomposition, please refer to Section 8.2.

The indices are first converted to a time series object with `ts()` and then are decomposed with function `stl()`. The decomposition below clearly shows an increase trend of HPI (see the 3rd chart in Figure 12.11).

```
> hpi <- ts(houseIndex$index, start=c(1990,1), frequency=12)
```

```
> f <- stl(hpi, "per")

> plot(f)
```

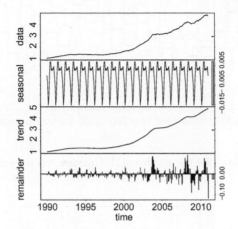

Figure 12.11 Decomposition of HPI data.

The code below produces a close look at the seasonal components (see Figure 12.12), which looks similar to the monthly average increased rates of HPI shown in Figure 12.7.

```
> # plot seasonal components

> plot(f$time.series[1:12,"seasonal"], type='b', xlab="Month",

+      ylab="Seasonal Components")
```

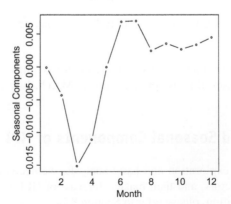

Figure 12.12 Seasonal components of HPI data.

An alternative function for decomposition is decompose() shown as below.

```
> # an alternative decomposition function
```

```
> f2 <- decompose(hpi)

> plot(f2)

> # plot seasonal components

> plot(f2$figure, type="b", xlab="Month", ylab="Seasonal
  Components")
```

12.4   HPI Forecasting

In this section, an ARIMA (autoregressive integrated moving average) model is fit to
HPI data and then used for forecasting HPIs in the next 4 years (see Figure 12.13). For
more details about time series forecasting, please refer to Section 8.3.

```
> startYear <- 1990

> endYear <- 2010

> # to forecast HPIs in the next four years

> nYearAhead <- 4

> fit <- arima(hpi, order=c(2,0,1), seasonal=list(order=c(2,1,0),
  period=12))

> fore <- predict(fit, n.ahead=12*nYearAhead)

> # error bounds at 95% confidence level

> U <- fore$pred + 2 * fore$se

> L <- fore$pred - 2 * fore$se

> # plot original and predicted data, as well as error bounds

> ts.plot(hpi, fore$pred, U, L, col=c("black",
  "blue","green","red"),

+          lty=c(1,5,2,2), gpars=list(xaxt="n",xlab=""),

+          ylab="Index", main="House Price Trading Index
  Forecast(Canberra)")

> # add labels, reference grid and legend

> years <- startYear:(endYear + nYearAhead+1)

> axis(1, labels=paste("Jan ", years, sep=""), las=3, at=years)
```

```
> grid()

> legend("topleft", col=c("black", "blue","green","red"),
  lty=c(1,5,2,2), c("Actual Index", "Forecast", "Upper Bound(95%
  Confidence)", "Lower Bound(95% Confidence)"))
```

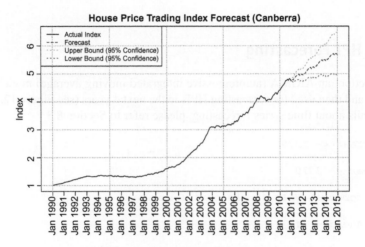

Figure 12.13 HPI forecasting—I.

To have a closer look at the forecasted CPIs, the forecasted values from 2011 are drawn below (see Figure 12.14).

```
> ts.plot(fore$pred, U, L, col=c("blue","green","red"),
  lty=c(5,2,2), gpars=list(xaxt="n",xlab=""), ylab="Index",
  main="House Price Trading Index Forecast(Canberra)")

> years <- endYear +(1:(nYearAhead+1))

> axis(1, labels=paste("Jan ", years, sep=""), las=3, at=years)

> grid(col = "gray", lty = "dotted")

> legend("topleft", col=c("blue","green","red"), lty=c(5,2,2),

+          c("Forecast", "Upper Bound (95% Confidence)",

+           "Lower Bound (95% Confidence)"))
```

Figure 12.14 HPI forecasting—II.

12.5   The Estimated Price of a Property

A property was sold at $535,000 in Canberra in September 2009, and what would be
its price two years later? The example below gives the estimated answer: $616,083.

```
> newHpi <- ts(c(hpi, fore$pred), start=c(1990,1), frequency=12)

> (startDate <- start(newHpi))

[1] 1990 1

> startYear <- startDate[1]

> m <- 9 + (2009-startYear)*12

> n <- 9 + (2011-startYear)*12

> # percentage of increase

> 100 * (newHpi[n]/ newHpi[m] - 1)

[1] 15.15576

> round(535000 * newHpi[n]/ newHpi[m])

[1] 616083
```

12.6   Discussion

This section presents a simple analysis on HPI data in a single city. The patterns of HPI
may be better studied by comparing the above HPI with HPI data in other cities, such

as Sydney and Melbourne. One may investigate whether they share similar patterns, whether there are any relationship between the changes of HPI in different cities, and whether there are any lags between major increases/decreases in different cities. Meanwhile, other factors, such as economic environment, population changes, CPI, and government policies, may also be included for a better forecasting of HPI with a regression model.

13 Case Study II: Customer Response Prediction and Profit Optimization

13.1 Introduction

In this case study, the competition of KDD Cup 1998[1] is used to demonstrate customer response prediction and profit maximization with decision trees. The same methodology has been applied successfully in a real business application, whose details, unfortunately, cannot be disclosed here due to the concern of customer privacy and business confidentiality.

The competition of the KDD Cup 1998 is to estimate the return from a direct mailing in order to maximize donation profits. To improve the efficiency of donation raising, we use data mining techniques to optimize customer selection. More specifically, decision trees are built with *R* to model donation raising based on customer demographics and promotion history. The objective is to predict the response of customers if contacted for the purpose of donation raising. By ranking customers based on predicted scores, the donation amount can be maximized.

The data mining process of this case study is shown in Figure 13.1 and we will describe these steps in the following sections.

13.2 The Data of KDD Cup 1998

The competition of KDD Cup 1998 is to estimate the return from a direct mailing in order to maximize the amount of donation. The datasets are in comma delimited format. The learning dataset "cup98LRN.txt" contains 95,412 records and 481 fields, and the validation dataset "cup98VAL.txt" contains 96,367 records and 479 variables. Each record has a field CONTROLN, which is a unique record identifier. There are two target variables in the learning dataset, TARGET_B and TARGET_D. TARGET_B is a binary variable indicating whether or not the record responded to mail while TARGET_D contains the donation amount in dollar. The learning dataset is of the same format as the validation one, except that the latter does not contain the above two target variables. The data can be downloaded at http://www.sigkdd.org/kddcup/index.php?section=1998&method=data.

[1] http://www.sigkdd.org/kddcup/index.php?section=1998&method=info

R and Data Mining. DOI: http://dx.doi.org/10.1016/B978-0-12-396963-7.00013-1

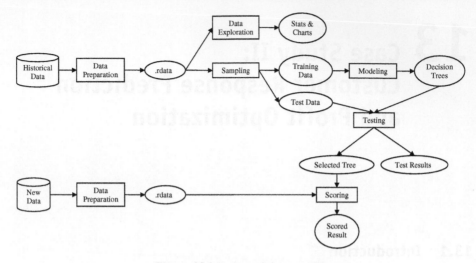

Figure 13.1 A data mining process.

Below we load the learning data into *R* and then have a look at it. To save space, we check only the first 30 variables with the code below.

```
> cup98 <- read.csv("./data/KDDCup1998/cup98LRN.txt")

> dim(cup98)

[1] 95412 481

> # have a look at the first 30 variables

> str(cup98[,1:30])

'data.frame':  95412 obs. of 30 variables:

$ ODATEDW: int 8901 9401 9001 8701 8601 9401 8701 9401 8801 9401 …

$ OSOURCE: Factor w/ 896 levels " ","AAA","AAD",..: 343 122 50 128
  1 220 255 613 487 549 …

$ TCODE  : int 0 1 1 0 0 0 0 0 1 1 …

$ STATE  : Factor w/ 57 levels "AA","AE","AK",..: 20 9 33 9 14 4
  21 24 18 48 …

$ ZIP    : Factor w/ 19938 levels "00801","00802",..: 9940 16858 336
  18629 2937 3841 5897 12146 7439 4251 …

$ MAILCODE: Factor w/ 2 levels " ","B": 1 1 1 1 1 1 1 1 1 1 …

$ PVASTATE: Factor w/ 3 levels " ","E","P": 1 1 1 1 1 1 1 1 1 1 …

$ DOB    : int 3712 5202 0 2801 2001 0 6001 0 0 3211 …
```

```
$ NOEXCH : Factor w/ 4 levels " ","0","1","X": 2 2 2 2 2 2 2 2
  2 …

$ RECINHSE: Factor w/ 2 levels " ","X": 1 1 1 1 2 1 1 1 1 …

$ RECP3 : Factor w/ 2 levels " ","X": 1 1 1 1 2 1 1 1 1 …

$ RECPGVG: Factor w/ 2 levels " ","X": 1 1 1 1 1 1 1 1 1 …

$ RECSWEEP: Factor w/ 2 levels " ","X": 1 1 1 1 1 1 1 1 1 …

$ MDMAUD : Factor w/ 28 levels "C1CM","C1LM",..: 28 28 28 28 28
  28 28 28 28 28 …

$ DOMAIN : Factor w/ 17 levels " ","C1","C2",..: 12 8 6 6 9 12 12
  12 6 11 …

$ CLUSTER: int 36 14 43 44 16 40 40 39 45 35 …

$ AGE    : int 60 46 NA 70 78 NA 38 NA NA 65 …

$ AGEFLAG: Factor w/ 3 levels " ","E","I": 1 2 1 2 2 1 2 1 1 3 …

$ HOMEOWNR: Factor w/ 3 levels " ","H","U": 1 2 3 3 2 1 2 3 3 1 …

$ CHILD03: Factor w/ 4 levels " ","B","F","M": 1 1 1 1 1 1 1 1 1
  1 …

$ CHILD07: Factor w/ 4 levels " ","B","F","M": 1 1 1 1 1 1 1 1 1
  1 …

$ CHILD12: Factor w/ 4 levels " ","B","F","M": 1 1 1 1 1 1 3 1 1
  1 …

$ CHILD18: Factor w/ 4 levels " ","B","F","M": 1 4 1 1 1 1 1 1 1
  1 …

$ NUMCHLD: int NA 1 NA NA 1 NA 1 NA NA NA …

$ INCOME : int NA 6 3 1 3 NA 4 2 3 NA …

$ GENDER : Factor w/ 7 levels " ","A","C","F",..: 4 6 6 4 4 1 4 4
  6 6 …

$ WEALTH1: int NA 9 1 4 2 NA 6 9 2 NA …

$ HIT    : int 0 16 2 2 60 0 0 1 0 0 …

$ MBCRAFT: int NA 0 0 0 1 NA NA 0 NA NA …

$ MBGARDEN: int NA 0 0 0 0 NA NA 0 NA NA …

> head(cup98[,1:30])
```

	ODATEDW	OSOURCE	TCODE	STATE	ZIP	MAILCODE	PVASTATE	DOB	NOEXCH	RECINHSE
1	8901	GRI	0	IL	61081			3712	0	
2	9401	BOA	1	CA	91326			5202	0	
3	9001	AMH	1	NC	27017			0	0	
4	8701	BRY	0	CA	95953			2801	0	
5	8601		0	FL	33176			2001	0	X
6	9401	CWR	0	AL	35603			0	0	

	RECP3	RECPGVG	RECSWEEP	MDMAUD	DOMAIN	CLUSTER	AGE	AGEFLAG	HOMEOWNR	CHILD03
1				XXXX	T2	36	60			
2				XXXX	S1	14	46	E	H	
3				XXXX	R2	43	NA		U	
4				XXXX	R2	44	70	E	U	
5	X			XXXX	S2	16	78	E	H	
6				XXXX	T2	40	NA			

	CHILD07	CHILD12	CHILD18	NUMCHLD	INCOME	GENDER	WEALTH1	HIT	MBCRAFT	MBGARDEN
1				NA	NA	F	NA	0	NA	NA
2		M		1	6	M	9	16	0	0
3				NA	3	M	1	2	0	0
4				NA	1	F	4	2	0	0
5				1	3	F	2	60	1	0
6				NA	NA		NA	0	NA	NA

```
> # a summary of the first 10 variables

> summary(cup98[,1:10])
```

```
    ODATEDW          OSOURCE           TCODE             STATE
 Min.   :8306    MBC    :4539   Min.   :    0.00    CA   :17343
 1st Qu.:8801    SYN    :3563   1st Qu.:    0.00    FL   :8376
 Median :9201    AML    :3430   Median :    1.00    TX   :7535
 Mean   :9141    BHG    :3324   Mean   :   54.22    IL   :6420
 3rd Qu.:9501    IMP    :2986   3rd Qu.:    2.00    MI   :5654
 Max.   :9701    ARG    :2409   Max.   :72002.00    NC   :4160
                 (Other):75161                      (Other):45924
      ZIP         MAILCODE       PVASTATE        DOB          NOEXCH      RECINHSE
 85351  :61                :94013  :93954   Min.   :   0        :    7   : 88709
 92653  :59    B    : 1399    E :    5   1st Qu. : 201   0: 95085   X:  6703
 85710  :54                   P : 1453   Median :2610   1:   285
 95608  :50                              Mean   :2724   X:    35
 60619  :45                              3rd Qu.:4601
 89117  :45                              Max.   :9710
 (Other):95098
```

To check all the variables in the data, readers can use the code below. The results are not included in this book to save space.

```
> library(Hmisc)

> describe(cup98[,1:28]) # demographics

> describe(cup98[,29:42]) # number of times responded to other
  types of mail order offers

> describe(cup98[,43:55]) # overlay data

> describe(cup98[,56:74]) # donor interests

> describe(cup98[,75]) # PEP star RFA status

> describe(cup98[,76:361]) # characteristics of the donors
  neighborhood

> describe(cup98[,362:407])# promotion history

> describe(cup98[,408:412])# summary variables of promotion
  history

> describe(cup98[,413:456])# giving history

> describe(cup98[,457:469])# summary variables of giving history

> describe(cup98[,470:473])# ID & targets

> describe(cup98[,474:479])# RFA (Recency/Frequency/Donation
  Amount)

> describe(cup98[,480:481])# CLUSTER & GEOCODE
```

Then we check the distribution of the two target variables, TARGET_B and TARGET_D. A pie chart of TARGET_B is shown in Figure 13.2.

```
> (response.percentage <- round(100 * prop.table(table(cup98$
  TARGET_B)), digits=1))

0     1
94.9  5.1


> mylabels <- paste("TARGET_B=", names(response.percentage), "\n",

+                    response.percentage, "%", sep=" ")

> pie(response.percentage, labels=mylabels)
```
(see Figure 13.2)

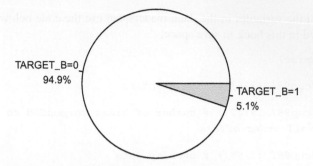

Figure 13.2 Distribution of response.

With the code below, we check positive donations, that is, those records with
TARGET_D greater than zero.

```
> # data with positive donations

> cup98pos <- cup98[cup98$TARGET_D >0,]

> targetPos <- cup98pos$TARGET_D

> summary(targetPos)
```

```
Min.   1st Qu.   Median   Mean   3rd Qu.   Max.
1.00   10.00     13.00    15.62  20.00     200.00
```

```
> boxplot(targetPos)
```
(see Figure 13.3)

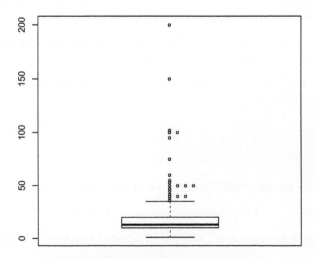

Figure 13.3 Box plot of donation amount.

Below we check the number of positive donations not in whole dollars, round the donation amount to whole dollars and then draw a barplot for it. The plot (see Figure 13.4) shows that most donations are no more than $25 and are multiples of $5.

```
> # number of positive donations

> length(targetPos)

[1] 4843

> # number of positive donations not in whole dollars

> sum(!(targetPos %in% 1:200))

[1] 21

> targetPos <- round(targetPos)

> barplot(table(targetPos), las=2)
```

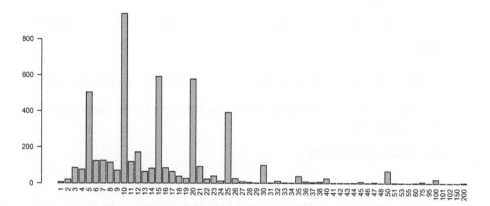

Figure 13.4 Barplot of donation amount.

Based on the distribution shown in the above barplot, we discretize TARGET_D to make a new variable TARGET_D2. Function cut( ) is used for discretization, where right=F indicates that the intervals are open on the right and closed on the left.

```
> cup98$TARGET_D2 <- cut(cup98$TARGET_D, right=F,
  breaks=c(0, 0.1, 10, 15, 20, 25, 30, 50, max(cup98$TARGET_D)))

> table(cup98$TARGET_D2)
```

[0,0.1)	[0.1,10)	[10,15)	[15,20)	[20,25)	[25,30)	[30,50)	[50,200)
90569	1132	1378	806	745	435	233	110

```
> cup98pos$TARGET_D2 <- cut(cup98pos$TARGET_D, right=F,
  breaks=c(0, 0.1, 10, 15, 20, 25, 30, 50, max(cup98pos$TARGET_D)))
```

Variable RFA_2R (recency code for RFA_2) is removed, because all records have
the same value of "L" in that field. Around 99.7% of records has a value of "0" in field
NOEXCH, so it is also removed.

```
> table(cup98$RFA_2R)


    L
95412
```

```
> round(100 * prop.table(table(cup98$NOEXCH)), digits=3)


        0        1        X
  0.007   99.657    0.299    0.037
```

We also excluded characteristics of the donors neighborhood, and variables from
the promotion history file and the giving history file. However, the summary variables
from the two history files are kept.

After the above inspection and exploration, the following variables are selected.

- Demographics:

  - ODATEDW: origin date. Date of donor's first gift in YYMM format (Year/Month);

  - OSOURCE: origin source. Code indicating which mailing list the donor was orig-
    inally acquired from;

  - STATE: state abbreviation;

  - ZIP: zipcode;

  - PVASTATE: indicates whether the donor lives in a state served by the organiza-
    tion's chapter;

  - DOB: date of birth (YYMM, Year/Month format.);

  - RECINHSE: in house file flag;

  - MDMAUD: the major donor matrix code. The codes describe frequency and amount
    of giving for donors who have given a $100+ gift at any time in their giving
    history. First byte: recency of giving; 2nd byte: frequency of giving; 3rd byte:
    amount of giving; 4th byte: blank/meaningless/filler.

  - DOMAIN: domain/cluster code. 1st byte = urbanicity level of the donor's neigh-
    borhood; 2nd byte = socio-economic status of the neighborhood.

  - CLUSTER: code indicating which cluster group the donor falls into. Each cluster
    is unique in terms of socio-economic status, urbanicity, ethnicity, and a variety
    of other demographic characteristics.

- AGE: overlay age;

- HOMEOWNR: home owner flag;

- CHILD03, CHILD07, CHILD12, CHILD18: presence of children age 0–3, 4–7, 8–12, and 13–18;

- NUMCHLD: number of children;

- INCOME: household income;

- GENDER: gender;

- WEALTH1: wealth rating;

- HIT: number of mail order responses;

- Donor interests: COLLECT1, VETERANS, BIBLE, CATLG, HOMEE, PETS, CDPLAY, STEREO, PCOWNERS, PHOTO, CRAFTS, FISHER, GARDENIN, BOATS, WALKER, KIDSTUFF, CARDS, PLATES;

• History information:

- PEPSTRFL: PEP star RFA status;

- summary variables of promotion history: CARDPROM, MAXADATE, NUMPROM, CARDPM12, NUMPRM12:

- summary variables of giving history: RAMNTALL, NGIFTALL, CARDGIFT, MINRAMNT, MAXRAMNT, LASTGIFT, LASTDATE, FISTDATE, TIMELAG, AVGGIFT;

• ID & targets:

- ID: CONTROLN;

- Targets: TARGET_D, TARGET_D2, TARGET_B;

• Others:

- Presence of published home phone number: HPHONE_D;

- RFA (Recency/Frequency/Donation Amount): RFA_2F, RFA_2A, MDMAUD_R, MDMAUD_F, MDMAUD_A; and

- Others codes: CLUSTER2, GEOCODE2.

```
> varSet <- c(

+    # demographics
```

```
+    "ODATEDW", "OSOURCE", "STATE", "ZIP", "PVASTATE", "DOB",
     "RECINHSE", "MDMAUD", "DOMAIN", "CLUSTER", "AGE", "HOMEOWNR",
     "CHILD03", "CHILD07", "CHILD12", "CHILD18", "NUMCHLD",
     "INCOME", "GENDER", "WEALTH1", "HIT",

+    # donor interests

+    "COLLECT1", "VETERANS", "BIBLE", "CATLG", "HOMEE", "PETS",
     "CDPLAY", "STEREO", "PCOWNERS", "PHOTO", "CRAFTS", "FISHER",
     "GARDENIN", "BOATS", "WALKER", "KIDSTUFF", "CARDS", "PLATES",

+    # PEP star RFA status

+    "PEPSTRFL",

+    # summary variables of promotion history

+    "CARDPROM", "MAXADATE", "NUMPROM", "CARDPM12", "NUMPRM12",

+    # summary variables of giving history

+    "RAMNTALL", "NGIFTALL", "CARDGIFT", "MINRAMNT", "MAXRAMNT",
     "LASTGIFT", "LASTDATE", "FISTDATE", "TIMELAG", "AVGGIFT",

+    # ID & targets

+    "CONTROLN", "TARGET_B", "TARGET_D", "TARGET_D2", "HPHONE_D",

+    # RFA (Recency/Frequency/Donation Amount)

+    "RFA_2F", "RFA_2A", "MDMAUD_R", "MDMAUD_F", "MDMAUD_A",

+    #others

+    "CLUSTER2", "GEOCODE2")
> cup98 <- cup98[, varSet]
```

13.3   Data Exploration

Generally speaking, data need to be explored in three steps. The first step is to check the distribution of individual variables. This is to know the distribution of values of each variable and check for missing values and outliers, so that we can know whether the variables need any transformations and whether they should be included in or excluded from modeling. The second step is to check the relationship between targets (dependent variables) and predictors (independent variables), which can be used for feature selection. The third step is to check the relationship among predictors themselves, so that redundant variables can be removed.

We first have a look at summary status of the data and the distribution of numeric variables. The code below can be easily changed to output plots of all variables to a PDF file. Examples of results are shown in Figures 13.5 and 13.6.

```
> # select numeric variables

> idx.num <- which(sapply(cup98, is.numeric))

> layout(matrix(c(1,2),1,2)) # 2 graphs per page

> # histograms of numeric variables

> myHist <- function(x) {

+    hist(cup98[,x], main=NULL, xlab=x)

+ }

> sapply(names(idx.num[4:5]), myHist)

             AGE              NUMCHLD
breaks       Numeric,21       Numeric,13
counts       Integer,20       Integer,12
intensities  Numeric,20       Numeric,12
density      Numeric,20       Numeric,12
mids         Numeric,20       Numeric,12
xname        "cup98[, x]"     "cup98[, x]"
equidist     TRUE             TRUE

> layout(matrix(1)) # change back to one graph per page
```

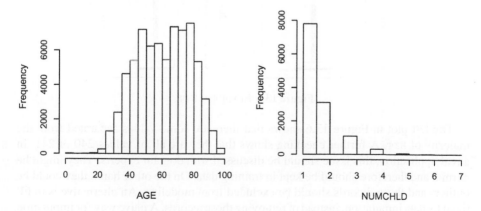

Figure 13.5 Histograms of numeric variables.

```
> # run code below to generate histograms for all numeric
  variables
```

```
> # sapply(names(idx.num), myHist)

> layout(matrix(c(1,2),1,2)) # 2 graphs per page

> boxplot(cup98$HIT)

> cup98$HIT[cup98$HIT>200]

[1]   240 240 240 240 240 240 240 240 240 240 240 240 240 240 240 240 240 240 240
[20]  240 240 240 240 240 241 240 240 240 240 241 240 240 240 240 240 240 240 240
[39]  240 240 240 240 240 240 240 240 240 240 240 240 240 240 241 241 240 240 240
[58]  240 240 240 240 240 240 240 240 240 240 240 240 240 240 240 240
```

```
> boxplot(cup98$HIT[cup98$HIT<200])

> layout(matrix(1)) # change back to one graph per page
```

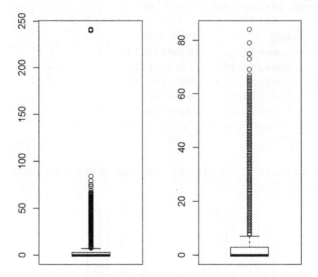

Figure 13.6 Boxplot of HIT.

The left plot in Figure 13.6 shows that there are some values separated from the majority of HIT. A further checking shows that they are all of values 240 or 241. In a real application, this issue should be discussed with domain experts. They might be normal and therefore should be kept in training data. On the other hand, they could be outliers and those records should be excluded from modeling. An alternative is to fill them by data imputation, instead of removing those records. A naive way for imputation is replacing the values with the mean or median of HIT in all records. In this exercise, the data are not imputed.

We then check the distribution of donation in various age groups. Figure 13.7 shows that people aged 30 to 60 are of higher median donation amount than others. It makes sense because they are the working force.

```
> AGE2 <- cut(cup98pos$AGE, right=F, breaks=seq(0, 100, by=5))

> boxplot(cup98pos$TARGET_D ~ AGE2, ylim=c(0,40), las=3)
```

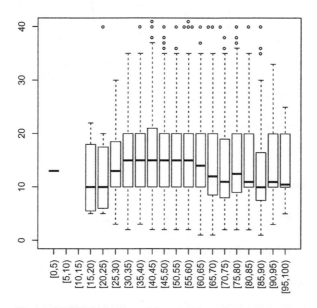

Figure 13.7 Distribution of donation in various age groups.

Below we check the distribution of donation amount for different genders. The results in Figure 13.8 show that the donation amount from joint account ("J") is less than male ("M") or female ("F").

```
> attach(cup98pos)

> layout(matrix(c(1,2),1,2)) # 2 graphs per page

> boxplot(TARGET_D GENDER, ylim=c(0,80))

> # density plot

> plot(density(TARGET_D[GENDER=="F"]), xlim=c(0,60), col=1, lty=1)

> lines(density(TARGET_D[GENDER=="M"]), col=2, lty=2)

> lines(density(TARGET_D[GENDER=="J"]), col=3, lty=3)

> legend("topright", c("Female", "Male", "Joint account"),
  col=1:3, lty=1:3)

> layout(matrix(1)) # change back to one graph per page

> detach(cup98pos)
```

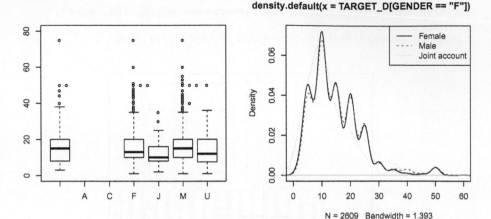

Figure 13.8 Distribution of donation in various age groups.

After that, we check the correlation between the target variable and other numeric variables with function `cor()`. By setting `use` to `pairwise.complete.obs`, the correlation between each pair of variables is computed using all complete pairs of observations on those variables, so that the resulting values will not be NA when there are missing values in the data.

```
> correlation <- cor(cup98$TARGET_D, cup98[,idx.num],
  use="pairwise.complete.obs")

> correlation <- abs(correlation)

> (correlation <- correlation[,order(correlation, decreasing=T)])

      TARGET_D      TARGET_B      LASTGIFT      RAMNTALL       AVGGIFT      MAXRAMNT
  1.0000000000  0.7742323755  0.0616784458  0.0448101061  0.0442990841  0.0392237509
        INCOME      CLUSTER2      NUMPRM12       WEALTH1      MINRAMNT      LASTDATE
  0.0320627023  0.0290870830  0.0251337775  0.0248673117  0.0201578686  0.0188471021
       NUMPROM       CLUSTER      CARDPM12       NUMCHLD      CONTROLN      CARDPROM
  0.0173371740  0.0171274879  0.0163577542  0.0149204899  0.0133664439  0.0113023931
      FISTDATE       ODATEDW           HIT      CARDGIFT      NGIFTALL      MAXADATE
  0.0075324932  0.0069484311  0.0066483728  0.0064498822  0.0048990126  0.0044963520
       TIMELAG           DOB      HPHONE_D           AGE        RFA_2F
  0.0036115917  0.0027541472  0.0024315898  0.0022823598  0.0009047682

> # save to a CSV file, with important variables at the top

> write.csv(correlation, "absolute_correlation.csv")
```

We also check the correlation between every pair of numeric variables and the scatter plot of every pair of variables.

```
> cor(cup98[,idx.num])

> pairs(cup98)
```

We then plotted scatter plots of numeric variables, with points colored based on target variables. Function `jitter()` is used below to introduce a small amount of noise, which is useful when there are many overlapping points. A scatter plot of AGE and HIT is shown in Figure 13.9.

```
> color <- ifelse(cup98$TARGET_D>0, "blue", "black")

> pch <- ifelse(cup98$TARGET_D>0, "+", ".")

> plot(jitter (cup98$AGE), jitter(cup98$HIT), pch=pch, col=color,
  cex=0.7, ylim=c(0,70), xlab="AGE", ylab="HIT")

> legend("topleft", c("TARGET_D>0", "TARGET_D=0"), col=c("blue",
  "black"), pch=c("+", "."))
```

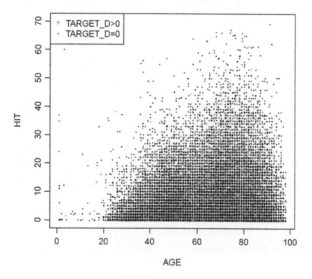

Figure 13.9 Scatter plot.

For categorical variables, we checked their association with chi-square test below (see Figure 13.10).

```
> myChisqTest <- function(x) {

+       t1 <- table(cup98pos[,x], cup98pos$TARGET_D2)

+       plot(t1, main=x, las=1)

+       print(x)

+       print(chisq.test(t1))

+ }

> myChisqTest("GENDER") (see Figure 13.10)
```

```
[1] "GENDER"

        Pearson's Chi-squared test

data: t1

X-squared=NaN, df=42, p-value=NA
```

GENDER

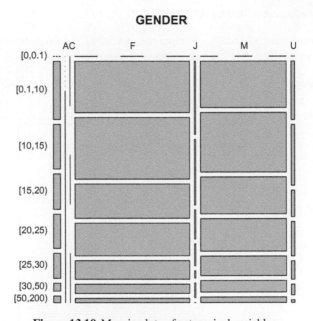

Figure 13.10 Mosaic plots of categorical variables.

```
> # run the code below to do chi-square test for all categorical
  variables

> # idx.cat <- which(sapply(cup98pos, is.factor))

> # sapply(names(idx.cat), myChisqTest)
```

13.4   Training Decision Trees

In this section, we build decision trees using R, with the ctree() function in package *party* (Hothorn et al. [2010]). There are a few parameters, MinSplit, MinBusket, MaxSurrogate, and MaxDepth, to control the training of decision trees. MinSplit is the minimum number of instances in a node in order to be considered for splitting, MinBusket sets the minimum number of instances in a terminal node, MaxSurrogate stands for the number of surrogate splits to evaluate, and MaxDepth controls the maximum depth of the tree.

With the code below, we set the sizes of training data (70%) and test data (30%), and the parameters for training decision trees. The MinSplit or MinBusket can be set to be of the same scale as 1/100 of training data. This parameter and others should be set based on business problem, domain expert's experience, data, the reasonable time to run an algorithm, and the expected size of trees.

```
> nRec <- dim(cup98)[1]

> trainSize <- round(nRec * 0.7)

> testSize <- nRec - trainSize

> # ctree parameters

> MinSplit <- 1000

> MinBucket <- 400

> MaxSurrogate <- 4

> MaxDepth <- 10

> (strParameters <- paste(MinSplit, MinBucket, MaxSurrogate,
  MaxDepth, sep="-"))

  [1] "1000-400-4-10"

> LoopNum <- 9

> # The cost for each contact is $0.68.

> cost <- 0.68

> varSet2 <- c("AGE", "AVGGIFT", "CARDGIFT", "CARDPM12",
  "CARDPROM", "CLUSTER2", "DOMAIN", "GENDER", "GEOCODE2", "HIT",
  "HOMEOWNR", "HPHONE_D", "INCOME", "LASTGIFT", "MAXRAMNT",
  "MDMAUD_F", "MDMAUD_R", "MINRAMNT", "NGIFTALL", "NUMPRM12",
  "PCOWNERS", "PEPSTRFL", "PETS", "RAMNTALL", "RECINHSE",
  "RFA_2A", "RFA_2F", "STATE", "TIMELAG")

> cup98 <- cup98[, c("TARGET_D", varSet2)]

> library(party) # for ctree
```

The data are partitioned with random sampling into training and test sets. With a single run, the built tree and tested result can be to some degree dependent on the partitioning of data. Therefore, for each set of parameters, we run partitioning, training, and testing for nine times and then use the average result to compare trees built with different parameters. In the code below, with function pdf(), we set the width and height of the graphics region and the point size to make a large tree fit nicely in an A4 paper. Function cumsum() calculates cumulative sum of a sequence of numbers.

```
> pdf(paste("evaluation-tree-", strParameters, ".pdf", sep=""),

+       width=12, height=9, paper="a4r", pointsize=6)

> cat(date(), "\n")

> cat(" trainSize=", trainSize, ", testSize=", testSize, "\n")
```

```
> cat(" MinSplit=", MinSplit, ", MinBucket=", MinBucket,

+       ", MaxSurrogate=",MaxSurrogate, ", MaxDepth=",MaxDepth, "\n\n")

> # run for multiple times and get the average result

> allTotalDonation <- matrix(0, nrow=testSize, ncol=LoopNum)

> allAvgDonation <- matrix(0, nrow=testSize, ncol=LoopNum)

> allDonationPercentile <- matrix (0, nrow=testSize, ncol=LoopNum)

> for (loopCnt in 1:LoopNum)

+       cat(date(), ": iteration = ", loopCnt, "\n")

+

+       # split into training data and testing data

+       trainIdx <- sample(1:nRec, trainSize)

+       trainData <- cup98[trainIdx,]

+       testData <- cup98[-trainIdx,]

+

+       # train a decision tree

+       myCtree <- ctree(TARGET_D ., data=trainData,
  controls=ctree_control(minsplit=MinSplit, minbucket=MinBucket,
  maxsurrogate=MaxSurrogate, maxdepth=MaxDepth))

+       # size of ctree

+       print(object.size(myCtree), units="auto")

+       save(myCtree, file=paste("cup98-ctree-", strParameters, "-run-",

+                       loopCnt, ".rdata", sep=""))

+

+       figTitle <- paste("Tree", loopCnt)

+       plot(myCtree, main=figTitle, type="simple",
  ip_args=list(pval=FALSE), ep_args=list(digits=0,abbreviate=TRUE),
  tp_args=list(digits=2))

+       #print(myCtree)

+

+       # test

+       pred <- predict(myCtree, newdata=testData)

+       plot(pred, testData$TARGET_D)
```

```
+       print(sum(testData$TARGET_D[pred > cost] - cost))

+       # quick sort is "unstable" for tie values, so it is used
  here to introduce a bit random for tie values

+       s1 <- sort(pred, decreasing=TRUE, method = "quick",
  index.return=TRUE)

+       totalDonation <- cumsum(testData$TARGET_D[s1$ix])#
  cumulative sum

+       avgDonation <- totalDonation / (1:testSize)

+       donationPercentile <- 100 * totalDonation /
  sum(testData$TARGET_D)

+       allTotalDonation[,loopCnt] <- totalDonation

+       allAvgDonation[,loopCnt] <- avgDonation

+       allDonationPercentile[,loopCnt] <- donationPercentile

+       plot(totalDonation, type="l")

+       grid()

+ }

> graphics.off()

> cat(date(), ": Loop completed.\n\n\n"

> fnlTotalDonation <- rowMeans(allTotalDonation)
```

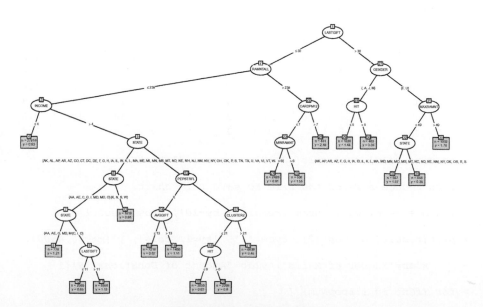

Figure 13.11 A decision tree.

```
> fnlAvgDonation <- rowMeans(allAvgDonation)

> fnlDonationPercentile <- rowMeans(allDonationPercentile)

> rm(trainData, testData, pred)

> # save results into a CSV file

> results <- data.frame(cbind(allTotalDonation,fnlTotalDonation))

> names(results) <- c(paste("run",1:LoopNum), "Average")

> write.csv(results, paste("evaluation-TotalDonation-",
  strParameters, ".csv", sep=""))
```

One of the built decision trees is shown in Figure 13.11.

13.5   Model Evaluation

With a decision tree model, the customers are ranked in descending order based on the predicted amount that they would donate. We plot the result of every run with the code below and the results are shown in Figures 13.12 and 13.13. In the figures the black solid line illustrates the average performance of all nine runs, while the other lines are the performance of individual runs. The two figures show that run 7 produced the best result.

With the code below, we draw a point for every ten points to reduce the size of files to save the charts. This is achieved with idx.pos.

```
> result <- read.csv("evaluation-TotalDonation-1000-400-4-10.csv")

> head(result)
```

	X	run.1	run.2	run.3	run.4	run.5	run.6	run.7	run.8	run.9	run.10	Average
1	1	0	0	0	0	0	0	0	0	18	0	1.8
2	2	0	0	0	0	0	0	0	0	18	0	1.8
3	3	0	0	0	0	0	0	0	0	18	0	1.8
4	4	0	0	0	0	0	0	0	0	18	10	2.8
5	5	0	0	0	0	0	0	0	0	18	10	2.8
6	6	0	0	0	50	0	0	0	0	18	10	7.8

```
> result[,2:12] <- result[,2:12] - cost * (1:testSize)

> # to reduce size of the file to save this chart

> idx.pos <- c(seq(1, nrow (result), by=10), nrow(result))

> plot(result[idx.pos,12], type="l", lty=1, col=1, ylim=c(0,4500),

+       xlab="Number of Mails", ylab="Amount of Donations ($)")

> for (fCnt in 1:LoopNum) {
```

```
+        lines (result[idx.pos,fCnt+1], pty=".", type="l",
  lty=1+fCnt, col=1+fCnt)

+}

> legend("bottomright", col=1:(LoopNum+1), lty=1:(LoopNum+1),

+        legend=c("Average", paste("Run",1:LoopNum)))
```

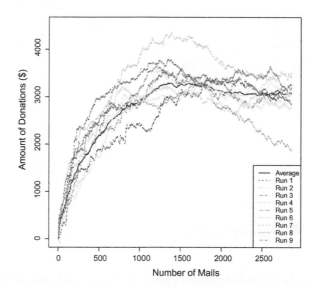

Figure 13.12 Total donation collected (1000–400–4–10).

```
> donationPercentile <- sapply(2:12, function(i)

+        100 * result[,i] / result[testSize,i])

> percentile <- 100 * (1:testSize)/testSize

> plot(percentile[idx.pos], donationPercentile[idx.pos,11],
  pty=".", type="l", lty=1, col=1, ylim=c(0,170), xlab="Contact
  Percentile (%)", ylab="Donation Percentile (%)")

> grid(col = "gray", lty = "dotted")

> for (fCnt in 1:LoopNum) {

+        lines(percentile[idx.pos], donationPercentile[idx.pos,fCnt],
  pty=".", type="l", lty=1+fCnt, col=1+fCnt)

+}
```

```
> legend("bottomright", col=1:(LoopNum+1), lty=1:(LoopNum+1),

+    legend=c("Average", paste("Run",1:LoopNum)))
```
(see Figure 13.13)

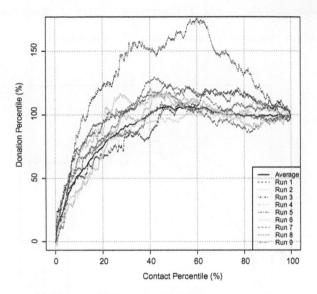

Figure 13.13 Total donation collected (9 runs).

Below we plot the evaluation result in a plot with double y-axis shown in Figure 13.14. It shows average result of the above nine runs, where the red solid line shows the percentage of donation amount collected and the blue dotted line shows the average donation amount by the customers contacted. The average donation amount per customer contacted is high in the left of the chart and then decreases when more customers are contacted. Therefore, the model is effective in capturing in its top-ranked list of the customers who would make big donations.

```
> avgDonation <- sapply(2:12, function(i) result[,i] /
  (1:testSize))

> yTitle = c("Total Donation Amount Percentile (%)",

+            "Average Donation Amount per Contact ($)")

> par(mar=c(5,4,4,5)+.1)

> plot(percentile[idx.pos], donationPercentile[idx.pos,7],
  pty=".", type="l", lty="solid", col="red", ylab=yTitle[1],
  xlab="Contact Percentile (%)")
```

```
> grid(col = "gray", lty = "dotted")

> par(new=TRUE)

> plot(percentile[idx.pos], avgDonation[idx.pos,7], type="l",
  lty="dashed", col="blue", xaxt="n", yaxt="n", xlab="", ylab="",
  ylim=c(0,max (avgDonation[,7])))

> axis(4)

> mtext(yTitle[2], side=4, line=2)

> legend("right", col=c("red","blue"), lty=c("solid","dashed"),

+          legend=yTitle)
```

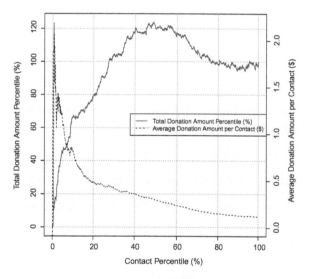

Figure 13.14 Average result of nine runs.

13.6   Selecting the Best Tree

We tested the decision trees generated with ctree() with six sets of different parameters. The average results of running each setting nine times are given in Figures 13.15 and 13.16. The labels in the legend show the values of MinSplit, MinBusket, MaxSurrogate, and MaxDepth used in the six sets of parameters. For example, with the first setting "1000–400–4–5", MinSplit is set to 1000, MinBusket is 400,

MaxSurrogate is 4, and MaxDepth is 5. Three different values are tested for MinSplit, which are 1000, 700, and 200. The corresponding values for MinBusket are 400, 200, and 50. The MaxDepth is also tried with four values: 5, 6, 8, and 10. The MaxSurrogate is set to 4 in all experiments.

Results are shown in Figures 13.15 and 13.16, where the horizontal axis represents the percentage of (ranked) customers contacted and the vertical axis shows the amount of donations that could be collected. A model is expected to collect more donations with the same number of contacts. The two figures are generated with the code below.

```
> # compare results got with different parameters

> parameters <- c("1000-400-4-5", "1000-400-4-6", "1000-400-4-8",
  "1000-400-4-10")

> #parameters <- c("1000-400-4-10", "700-200-4-10", "200-50-4-10")

> paraNum <- length(parameters)

> percentile <- 100 * (1:testSize)/testSize

> # 1st result

> results <- read.csv(paste("evaluation-TotalDonation-",
  parameters[1], ".csv", sep=""))

> avgResult <- results$Average - cost * (1:testSize)

> plot(percentile, avgResult, pty=1, type="l", lty=1, col=1,

+       ylab="Amount of Donation", xlab="Contact Percentile (%)",

+       main="Parameters: MinSplit, MinBucket, MaxSurrogate,
  MaxDepth")

>grid(col = "gray", lty = "dotted")

> # other results

> for (i in 2:paraNum) {

+       results <- read.csv(paste("evaluation-TotalDonation-",
  parameters[i], ".csv", sep=""))

+       avgResult <- results$Average - cost * (1:testSize)

+       lines(percentile, avgResult, type="l", lty=i, col=i)

+}

> legend("bottomright", col=1:paraNum, lty=1:paraNum,
  legend=parameters)
```

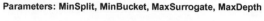

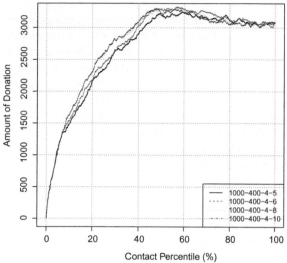

Figure 13.15 Comparison of different parameter settings—I.

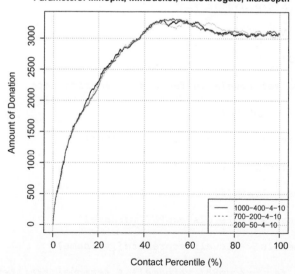

Figure 13.16 Comparison of different parameter settings—II.

Figure 13.15 shows that results with depth 8 and 10 are better than depth 5 and 6. Figure 13.16 shows that the three different sets of minimum bucket size and minimum split size have very similar results. We choose "1000–400–4–10" to produce the final model, because it is less likely to overfit than other models with smaller minimum bucket sizes and split sizes.

13.7   Scoring

Now we have trained a couple of decision trees with the learning dataset "cup98LRN.txt" and have selected the best one from them. Next, we will use the selected tree to score the validation dataset "cup98VAL.txt". People with a predicted donation amount greater than 0.68, the cost of contact, would be mailed for donation purpose. The evaluation criterion is the total amount of donations deducted by the total cost of mail.

Before scoring, we need to make sure that the data to score are of the same format as the training data used to build the model, which is essential for the prediction function to work. With the code below, we load the scoring data into R and set its factor levels to those in the training data. Unknown or new values in categorical variables in scoring data were set to NA (missing value).

```
> cup98val <- read.csv("./data/KDDCup1998/cup98VAL.txt")

> cup98val <- cup98val[, c("CONTROLN", varSet2)]

> trainNames <- names(cup98)

> scoreNames <- names(cup98val)

> # check if any variables not in scoring data

> idx <- which(trainNames %in% scoreNames)

> print(trainNames[-idx])

[1] "TARGET_D"

> # check and set levels in factors in scoring data

> scoreData <- cup98val

> vars <- intersect(trainNames, scoreNames)

> for (i in 1:length (vars)) {

+       varname <- vars[i]

+       trainLevels <- levels(cup98[,varname])

+       scoreLevels <- levels(scoreData[,varname])

+       if (is.factor(cup98[,varname]) & setequal(trainLevels,
  scoreLevels)==F) {

+            cat("Warning: new values found in score data, and they
  will be changed to NA!\n")

+            cat(varname, "\n")
```

```
+             #cat("train: ", length(trainLevels), ", ", trainLevels,
  "\n")

+             #cat("score: ", length (scoreLevels), ", ",
  scoreLevels,  "\n\n")

+             scoreData[,varname] <- factor(scoreData[,varname],
  levels=trainLevels)

+       }

+}
```

Warning: new values found in score data, and they will be changed
  to NA!

GENDER

Warning: new values found in score data, and they will be changed
  to NA!

STATE

```
> rm(cup98val)
```

After preparing the data to score, we then make predictions for them.

```
> # loading the selected model

> load("cup98-ctree-1000-400-4-10-run-7.Rdata")

> # predicting

> pred <- predict(myCtree, newdata=scoreData)

> pred <- round(pred, digits=3)

> #table(pred, useNA="ifany")

> result <- data.frame(scoreData$CONTROLN, pred)

> names(result) <- c("CONTROLN", "pred")

> valTarget <- read.csv("./data/KDDCup1998/valtargt.txt")

> merged <- merge(result, valTarget, by="CONTROLN")

> # donation profit if mail all people

> sum(valTarget$TARGET_D - cost)

[1] 10560.08
```

```
> # donation profit if mail those predicted to donate more than
  mail cost

> idx <- (merged$pred > cost)

> sum(merged$TARGET_D[idx] - cost)
```

[1] 13087.33

The above result shows that the model would produce a profit of $13,087, which would make it ranked no. 7 in the competition of KDD CUP 1998.

```
> # ranking customers

> merged <- merged[order (merged$pred, decreasing=T),]

> x <- 100 * (1:nrow (merged)) / nrow(merged)

> y <- cumsum(merged$TARGET_D) - cost*(1:nrow (valTarget))

> # to reduce size of the file to save this chart

> idx.pos <- c(seq(1, length (x), by=10), length(x))

> plot(x[idx.pos], y[idx.pos], type="l", xlab="Contact Percentile
  (%)", ylab="Amount of Donation")
```

> grid() (see Figure 13.17)

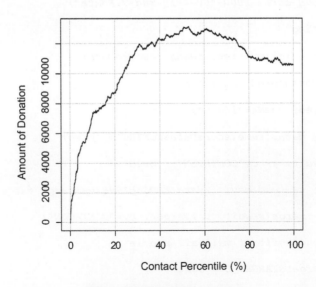

Figure 13.17 Validation result.

13.8 Discussions and Conclusions

This chapter presents a case study on profit optimization with decision trees. It clearly describes the process and data mining techniques, and provides R code examples, which readers can follow and apply in their own projects.

The aim of this chapter is to demonstrate how to build decision trees for real-world applications, and the built model is not the best model, even when compared with the models built for the KDD Cup in 1998. The readers are suggested to try the following methods to build better models.

The first method is to use a two-stage model, which was adopted by the gold winner of the KDD Cup 1998. With a two-stage model, the first model is to predict the probability of donation, the second model is to predict the conditional donation amount, and the product of the above two prediction produces an unconditional prediction of donation amount. More details about the method can be found at http://www.kdnuggets.com/meetings/kdd98/gain-kddcup98-release.html.

One might also try to make the data balanced. As shown in Figure 13.2, the percentage of people with donations is only 5.1% and the majority have not made any donations. The data can be made balanced by downsampling non-donated cases and/or oversampling donated ones, which might make it easier to build a predictive model and might produce better models.

Another method to try is derive new variables based on dates and historical donations. In this case study, no date variables or historical donations were used for modeling. Actually, some potentially useful information can be extracted from them, such as the number of days since last donation, and the number/amount of donations in the last one/two/three years. New derived variables may help to improve the performance of predictive models.

In this case study, some categorical variables with many levels were not included in the modeling process, because they would make the algorithms consume much more RAM and take much longer to run. However, they can be grouped to reduce the number of levels, especially for those infrequent levels, such as states and zipcodes with very small population.

One can also impute data by removing outliers and filling missing values, which were not covered in this case study.

14 Case Study III: Predictive Modeling of Big Data with Limited Memory

This chapter shows a case study on building a predictive model with limited memory. Because the training dataset was large and not easy to build decision trees within R, multiple subsets were drawn from it by random sampling, and a decision tree was built for each subset. After that, the variables appearing in any one of the built trees were used for variable selection from the original training dataset to reduce data size. In the scoring process, the scoring dataset was also split into subsets, so that the scoring could be done with limited memory. R codes for printing rules in plain English and in SAS format are also presented in this chapter.

14.1 Introduction

In this case study, we still tackle the problem of the KDD Cup 1998, which has been used in Chapter 13. The same methodology has been applied successfully in a real-world business application. However, the details of that application cannot be disclosed due to customer privacy and business confidentiality. Therefore, the data of the KDD Cup 1998 is used here to demonstrate the methodology for predictive modeling of big data with limited memory.

The data are in comma delimited format. The learning dataset "cup98LRN.txt" contains 95,412 records and 481 fields, and the validation dataset "cup98VAL.txt" contains 96,367 records and 479 variables. They contain many categorical variables, some of which have many value levels. A detailed description of the KDD Cup 98 datasets can be found in Section 13.2.

In this case study, we change the objective to predicting the likelihood that people make donations, that is, the target variable is set to TARGET_B, a binary variable indicating whether or not the record responded to mail. Note that in Chapter 13, TARGET_D, the donation amount in dollar, was used as the target variable. This change is to demonstrate how to predict probability, instead of a dollar amount.

The experiments in this chapter were conducted on a PC running Windows XP Professional SP3, with an Intel dual Core i5 3.1GHz CPU and 4 GB RAM. Although many PCs are much more powerful and have more RAM than the above one, the methodology presented in this chapter should be helpful when analyzing big data.

R and Data Mining. DOI: http://dx.doi.org/10.1016/B978-0-12-396963-7.00014-3

14.2   Methodology

In the data of this case study, there are two classes: target customers and non-target customers, labeled respectively as 1 and 0. It is similar with risk modeling of customers.

The technique of decision trees is used in this case study, because decision trees are easy to understand by business people and management, and the rules are simple and easy to be accepted and implemented by business, as compared to SVM or neural networks. They support mixed type data with both categorical and numerical variables, and can also handle missing values. Specifically, function ctree() in package *party* is used to build trees.

It took too long to train a model with large data, especially with some categorical variables having many value levels. An option is to use a small sample to train models. We took a different way by using as many training data as possible. To make it work, we draw 20 random samples of training data, and built 20 decision trees, one tree for each sample. There are around 20–30 variables in each tree, and many trees share similar set of variables. Then we collected all variables appearing in those trees, and got around 60 variables. After that, we used all original training data for training without any sampling, but with the above 60 variables only. In this way, all training cases were used to build a final model, but with only those attributes having appeared in the 20 trees built on sampled data.

14.3   Data and Variables

With the code below, the training data are loaded into R, and a set of variables are selected.

```
> cup98 <- read.csv("./data/KDDCup1998/cup98LRN.txt")

> dim(cup98)

[1] 95412 481

> n.missing <- rowSums(is.na (cup98))

> sum(n.missing > 0)

[1] 95412

> varSet <- c(

+   # demographics

+   "ODATEDW", "OSOURCE", "STATE", "ZIP", "PVASTATE", "DOB",
    "RECINHSE", "MDMAUD", "DOMAIN", "CLUSTER", "AGE", "HOMEOWNR",
    "CHILD03", "CHILD07", "CHILD12", "CHILD18", "NUMCHLD",
    "INCOME", "GENDER", "WEALTH1", "HIT",

+   # donor interests
```

```
+   "COLLECT1", "VETERANS", "BIBLE", "CATLG", "HOMEE", "PETS",
    "CDPLAY", "STEREO", "PCOWNERS", "PHOTO", "CRAFTS", "FISHER",
    "GARDENIN", "BOATS", "WALKER", "KIDSTUFF", "CARDS", "PLATES",

+   # PEP star RFA status

+   "PEPSTRFL",

+   # summary variables of promotion history

+   "CARDPROM", "MAXADATE", "NUMPROM", "CARDPM12", "NUMPRM12",

+   # summary variables of giving history

+   "RAMNTALL", "NGIFTALL", "CARDGIFT", "MINRAMNT", "MAXRAMNT",
    "LASTGIFT", "LASTDATE", "FISTDATE", "TIMELAG", "AVGGIFT",

+   # ID & targets

+   "CONTROLN", "TARGET_B", "TARGET_D", "HPHONE_D",

+   # RFA (Recency/Frequency/Donation Amount)

+   "RFA_2F", "RFA_2A", "MDMAUD_R", "MDMAUD_F", "MDMAUD_A",

+   #others

+   "CLUSTER2", "GEOCODE2")

> # remove ID & TARGET_D

> vars <- setdiff(varSet, c("CONTROLN", "TARGET_D"))

> cup98 <- cup98[,vars]
```

14.4 Random Forest

We first try to build random forests with two R packages, *randomForest* and *party*.

Package *randomForest* cannot handle missing values or categorical variables with more than 32 levels. Unfortunately, all records in the learning data have one or more missing values. Even in the data with variables in varSet only, there are about 93% of the records containing one or more missing values. It is a common situation in real-world data that most records have missing values. There are often categorical variables with more than 32 levels, such as country, ZIP code, occupation, and manufacturer. Some of them can be grouped into fewer categories, such as occupation. The levels of others can be reduced by putting levels with few records into groups, such as small countries and manufacturers.

```
> library(randomForest)

> rf <- randomForest(TARGET_B ~ ., data=cup98)
```

Below we check data for missing values and categorical variables with over ten levels.

```
> # check missing values

> n.missing <- rowSums(is.na(cup98))

> (tab.missing <- table(n.missing))

n.missing
0            1        2        3        4        5       6      7
6782         36864    23841    13684    11716    2483    41     1

> # percentage of records without missing values

> round(tab.missing["0"]/ nrow(cup98), digits=2)

0
0.07

> # check levels of categorical variables

> idx.cat <- which(sapply(cup98, is.factor))

> all.levels <- sapply(names(idx.cat), function(x)
  nlevels(cup98[,x]))

> all.levels[all.levels > 10]

OSOURCE    STATE    ZIP      MDMAUD    DOMAIN
896        57       19938    28        17
```

Below we split the data into training and test subsets.

```
> trainPercentage <- 80

> testPercentage <- 20

> ind <- sample(2, nrow(cup98), replace=TRUE,

+                prob=c(trainPercentage, testPercentage))

> trainData <- cup98[ind==1,]

> testData <- cup98[ind==2,]
```

We then try random forest with cforest() in package *party* as below. With 80% training data, it took about 2 minutes to build one tree and would take around 1.5 hours to build a random forest of 50 trees.

```
> # cforest

> library(party)

> (time1 <- Sys.time())
```

```
> cf <- cforest(TARGET_B~., data=trainData,

+          control = cforest_unbiased(mtry = 2, ntree = 50))

> (time2 <- Sys.time())

> time2 - time1

> print(object.size(cf), units = "Mb")

> myPrediction <- predict(cf, newdata=testData)

> (time3 <- Sys.time())

> time3 - time2
```

14.5 Memory Issue

In the rest of this chapter, we will build decision trees with function ctree() in package *party*.

```
> memory.limit(4095)

[1] 4095

> library(party)

> ct <- ctree(TARGET_B ~ ., data=trainData)
```

In the above code, memory.limit() sets the limit of memory (in MB) available to R. Another function memory.size() reports the current or maximum memory used by R. A useful function to check what memory is used for memory.profile(). Function object.size() returns the size of memory used by an R object. Details on memory allocation in R can be found by running ?memory.size.

When running the above code to build a decision tree with ctree(), we encountered a problem of memory. On a PC running Windows XP with 4GB RAM, we got an error message: "Error: cannot allocate vector of size 652.2 Mb", when the required memory is larger than 3GB on a 32-bit Windows machine. On a 64-bit Windows machine with 4GB RAM, it ran out of physical memory with the following error message:

```
Error: cannot allocate vector of size 3.0 Gb

In addition: Warning messages:

1: In as.vector(data):

Reached total allocation of 4095 Mb: see help(memory.size)

...
```

One way to reduce memory requirement is to group or remove categorical variables which have many levels. We first tried to use 20% of data for training, which contains around 19,200 rows and 62 columns. Function ctree() returned an error "reach total memory allocation" when ZIP was included. After removing ZIP, it ran successfully, but took 25 min. We then removed OSOURCE, and it built a decision tree in 5 seconds. We also tried to feed 80% of data (around 76,000 rows and 60 columns) into ctree() with both ZIP and OSOURCE removed and it completed in 25 seconds.

14.6   Train Models on Sample Data

To find out which variables are useful for modeling, the process in this section is repeated ten times to build ten decision trees. The variables appearing in any of the ten trees were collected and used to build a final model in the next section.

We first split the data into three subsets, training data (30%), test data (20%), and the rest. The reason for withholding some data as the rest is to reduce the size of training and test data, so that the training and test can be completed successfully on a machine with limited memory.

```
> library(party) # for ctree

> trainPercentage <- 30

> testPercentage <- 20

> restPrecentage <- 100 - trainPercentage - testPercentage

> fileName <- paste("cup98-ctree", trainPercentage,
  testPercentage, sep="-")

> vars <- setdiff(varSet, c("TARGET_D", "CONTROLN", "ZIP",
  "OSOURCE"))

> # partition the data into training and test datasets

> ind <- sample(3, nrow(cup98), replace=T,

+             prob=c(trainPercentage, testPercentage, restPrecentage))

> trainData <- cup98[ind==1, vars]

> testData <- cup98[ind==2, vars]
```

After sampling, we check whether the distribution of targets in both training and test data are the same as that in the original data. If not, stratified sampling might be used.

```
> # check the percentage of classes

> round(prop.table(table(cup98$TARGET_B)), digits=3)

0       1
0.949   0.051

> round(prop.table(table(trainData$TARGET_B)), digits=3)

0       1
0.949   0.052
```

```
> round(prop.table(table(testData$TARGET_B)), digits=3)

 0        1
 0.949    0.051

> # remove raw data to save memory

> rm(cup98, ind)

> gc()
          used      (Mb)    gc    trigger    (Mb)     max used    (Mb)
Ncells    536865    28.7          818163     43.7     741108      39.6
Vcells    2454644   18.8          20467682   156.2    78071440    595.7

> memory.size()
```

[1] 57.95

After that, we then use function `ctree()` to build a decision tree with the training data. To make the examples simple and easy to read, in this chapter, we use default settings when calling `ctree()` to train decision trees. For examples on setting the parameters for decision trees, please refer to Section 13.4. In the code below, function `object.size()` returns the size of a data object.

```
> # build ctree

> myCtree <- NULL

> startTime <- Sys.time()

> myCtree <- ctree(TARGET_B~., data=trainData)

> Sys.time() - startTime
```

Time difference of 8.802615 s

```
> print(object.size(myCtree), units = "Mb")
```

10.1 Mb

```
> #print(myCtree)

> memory.size()
```

[1] 370.7

```
> # plot the tree and save it in a .PDF file

> pdf(paste(fileName, ".pdf", sep=""), width=12, height=9,

+        paper="a4r", pointsize=6)

> plot(myCtree, type="simple", ip_args=list(pval=F),
  ep_args=list(digits=0), main=fileName)
```

```
> graphics.off()
```

The above process in this section is repeated ten times to build ten trees.

14.7   Build Models with Selected Variables

After building ten decision trees, variables appearing in any of the ten trees are collected and used for building a final model. At this time, all data are used for learning, with 80% for training and 20% for testing.

```
> vars.selected <- c("CARDS", "CARDGIFT", "CARDPM12", "CHILD12",
  "CLUSTER2", "DOMAIN", "GENDER", "GEOCODE2", "HIT", "HOMEOWNR",
  "INCOME", "LASTDATE", "MINRAMNT", "NGIFTALL", "PEPSTRFL",
  "RECINHSE", "RFA_2A", "RFA_2F", "STATE", "WALKER")
```

```
> trainPercentage <- 80
```

```
> testPercentage <- 20
```

```
> fileName <- paste("cup98-ctree", trainPercentage,
  testPercentage, sep="-")
```

```
> vars <- c("TARGET_B", vars.selected)
```

```
> # partition the data into training and test subsets
```

```
> ind <- sample(2, nrow(cup98), replace=T, prob=c(trainPercentage,
  testPercentage))
```

```
> trainData <- cup98[ind==1, vars]
```

```
> testData <- cup98[ind==2, vars]
```

```
> # build a decision tree
```

```
> myCtree <- ctree(TARGET_B~., data=trainData)
```

```
> print(object.size(myCtree), units = "Mb")
```

```
39.7 Mb
```

```
> memory.size()
```

```
[1] 1010.44
```

```
> print(myCtree)
```

```
        Conditional inference tree with 21 terminal nodes

Response: TARGET_B
Inputs: CARDS, CARDGIFT, CARDPM12, CHILD12, CLUSTER2, DOMAIN,
  GENDER, GEOCODE2, HIT, HOMEOWNR, INCOME, LASTDATE, MINRAMNT,
  NGIFTALL, PEPSTRFL, RECINHSE, RFA_2A, RFA_2F, STATE, WALKER
```

```
Number of observations: 76450

1) RFA_2A == {D, E}; criterion = 1, statistic = 428.147
  2) LASTDATE <= 9606; criterion = 1, statistic = 93.226
    3) RFA_2F <= 2; criterion = 1, statistic = 87.376
      4) INCOME <= 1; criterion = 0.985, statistic = 77.333
          5)* weights = 903
      4) INCOME > 1
          6)* weights = 6543
    3) RFA_2F > 2
      7) CARDPM12 <= 4; criterion = 1, statistic = 54.972
          8)* weights = 1408
      7) CARDPM12 > 4
        9) PEPSTRFL == {X}; criterion = 1, statistic = 47.597
          10) WALKER == {Y}; criterion = 1, statistic = 40.911
            11)* weights = 1152
          10) WALKER == {}
            12)* weights = 8479
        9) PEPSTRFL == {}
            13)* weights = 3804
      2) LASTDATE > 9606
        14)* weights = 1000
  1) RFA_2A == {F, G}
    15) PEPSTRFL == {X}; criterion = 1, statistic = 102.032
      16) LASTDATE <= 9607; criterion = 1, statistic = 48.418
        17) MINRAMNT <= 12.5; criterion = 1, statistic = 45.804
          18) RFA_2F <= 1; criterion = 1, statistic = 51.858
            19)* weights = 8121
          18) RFA_2F > 1
            20) GENDER == { , A, J, M, U}; criterion = 0.998,
statistic = 46.458
              21) GENDER == {A, J}; criterion = 0.998,
statistic = 37.321
                22)* weights = 38
              21) GENDER == { , M, U}
                23)* weights = 3591
            20) GENDER == {F}
              24)* weights = 4428
        17) MINRAMNT > 12.5
          25) CARDPM12 <= 4; criterion = 0.983,
statistic = 37.436
            26) NGIFTALL <= 2; criterion = 0.986,
statistic = 11.874
              27)* weights = 9
```

```
              26) NGIFTALL > 2
                 28)* weights = 39
            25) CARDPM12 > 4
               29)* weights = 605
      16) LASTDATE > 9607
         30) CARDPM12 <= 10; criterion = 1, statistic = 31.728
           31)* weights = 881
         30) CARDPM12 > 10
           32)* weights = 113
      15) PEPSTRFL == {}
         33) CARDGIFT <= 5; criterion = 1, statistic = 90.915
         34) CLUSTER2 <= 34; criterion = 1, statistic = 91.259
           35)* weights = 19613
         34) CLUSTER2 > 34
           36) RFA_2A == {F}; criterion = 0.966,
statistic = 58.501
               37)* weights = 10712
         36) RFA_2A == {G}
           38)* weights = 3843
   33) CARDGIFT > 5
       39) RFA_2F <= 2; criterion = 0.974, statistic = 39.703
       40)* weights = 951
       39) RFA_2F > 2
         41)* weights = 217
```

Then the built tree is saved into a Rdata file and the plot of it is saved into a PDF file. When a decision tree is big, the nodes and text in its plot may overlap with each other. A trick to avoid that is to set a big paper size (with `width` and `height`) and a small font (with `pointsize`). Moreover, the text in the plot can be reduced when plotting the tree, with `ip_args=list(pval=FALSE)` to suppress p-values and `ep_args=list (digits=0)` to reduce the length of numeric values. A plot of the tree is shown in Figure 14.1.

```
> save(myCtree, file = paste (fileName, ".Rdata", sep=""))

> pdf(paste(fileName, ".pdf", sep=""), width=12, height=9,

+   paper="a4r", pointsize=6)

> plot(myCtree, type="simple", ip_args=list(pval=F), ep_args=list
  (digits=0), main=fileName)

> plot(myCtree, terminal_panel=node_barplot(myCtree),
  ip_args=list(pval=F), ep_args=list(digits=0), main=fileName)

> graphics.off()
```

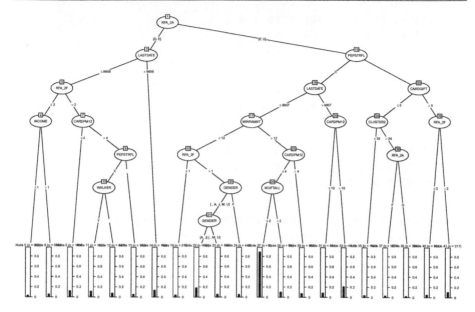

Figure 14.1 Decision tree.

The built model is then tested with the test data, and the test results are checked (see Figures 14.2, 14.3 and 14.4).

```
> rm(trainData)

> myPrediction <- predict(myCtree, newdata=testData)

> # check predicted results

> testResult <- table(myPrediction, testData$TARGET_B)

> percentageOfOne <- round(100 * testResult[,2] / (testResult[,1]
  + testResult[,2]), digits=1)

> testResult <- cbind(testResult, percentageOfOne)

> print(testResult)

                      0      1    percentageOfOne
0.0223783502472027   884    23    2.5
0.0310077519379845   260     8    3.0
0.0323935772964899  2541    82    3.1
0.0377810635802784  4665   214    4.4
0.0426055904445265  2007    75    3.6
0.0525762355415352   208    10    4.6
0.0535230352303523  1046    54    4.9
0.0557308096740273   841    50    5.6
0.0570074889194559  1587    90    5.4
```

```
0.0573656363130047    845    55    6.1
0.0743801652892562    160    9     5.3
0.0764241066163463    1895   154   7.5
0.0851305334846765    204    10    4.7
0.102564102564103     15     2     11.8
0.105990783410138     45     6     11.8
0.112847222222222     232    28    10.8
0.122159090909091     309    42    12.0
0.135                 240    26    9.8
0.184210526315789     8      0     0.0
0.212389380530973     22     8     26.7
0.888888888888889     2      0     0.0
```

```
> boxplot(myPrediction ~ testData$TARGET_B, xlab="TARGET_B",
  ylab="Prediction", ylim=c(0,0.25))
```

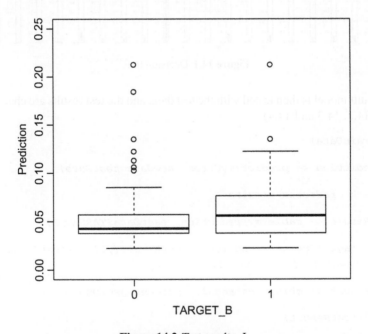

Figure 14.2 Test result—I.

```
> s1 <- sort(myPrediction, decreasing=TRUE, method = "quick",
  index.return=TRUE)
```

```
> testSize <- nrow(testData)
```

```
> TotalNumOfTarget <- sum(testData$TARGET_B)
```

```
> NumOfTarget <- rep(0, testSize)
```

```
> NumOfTarget[1] <- (testData$TARGET_B)[s1$ix[1]]

> for (i in 2:testSize) {

+  NumOfTarget[i] <- NumOfTarget[i-1] + testData$TARGET_B[s1$ix[i]]

+ }

> plot(1:testSize, NumOfTarget, pty=".", type="l", lty="solid",
  col="red", ylab="Count Of Responses in Top k", xlab="Top k",
  main=fileName)

> grid(col = "gray", lty = "dotted")
```

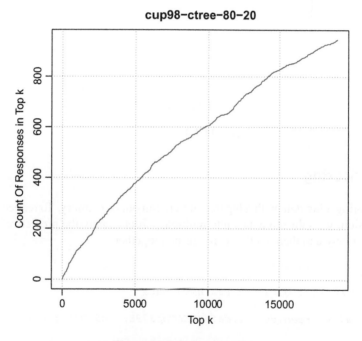

Figure 14.3 Test result—II.

```
> percentile <- 100 * (1:testSize)/ testSize

> percentileTarget <- 100 * NumOfTarget/ TotalNumOfTarget

> plot(percentile, percentileTarget, pty=".", type="l",
  lty="solid", col="red", ylab="Percentage of Predicted Donations
  (%)", xlab="Percentage of Pool", main=fileName)

> grid(col = "gray", lty = "dotted")
```

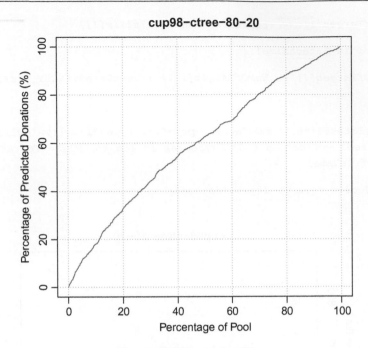

Figure 14.4 Test result—III.

14.8 Scoring

When scoring a big data with a big tree, it may run out of memory. To reduce memory consumption, we split score data into multiple subsets, apply the predictive model to them separately and then put the scored results together.

```
> memory.limit(4095)

> # read scoring data and training data

> cup98val <- read.csv("./data/KDDCup1998/cup98VAL.txt")

> cup98 <- read.csv("./data/KDDCup1998/cup98LRN.txt")

> library(party) # for ctree

> treeFileName <- "cup98-ctree-80-20"

> splitNum <- 10
```

Before scoring, we need to check whether the categorical variables in scoreData are of the same levels as those in trainData. If not, we need to set factor levels in scoreData to those in trainData, which is essential for the predict() function to work. Unknown or new values in categorical variables in scoring data are set to NAs (i.e., missing values).

```
> # check and set levels of categorical variables

> trainData <- cup98[,vars]

> vars2 <- setdiff(c(vars,"CONTROLN"), "TARGET_B")

> scoreData <- cup98val[,vars2]

> rm(cup98, cup98val)

> trainNames <- names(trainData)

> scoreNames <- names(scoreData)

> #cat("\n checking and setting variable values \n")

> newScoreData <- scoreData

> variableList <- intersect(trainNames, scoreNames)

> for (i in 1:length(variableList)) {

+   varname <- variableList[i]

+   trainLevels <- levels(trainData[,varname])

+   scoreLevels <- levels(newScoreData[,varname])

+   if (is.factor(trainData[,varname]) & setequal(trainLevels,
      scoreLevels)==F) {

+       cat("Warning: new values found in score data, and they
          will be changed to NA!\n")

+       cat(varname, "\n")

+       cat("train: ", length(trainLevels), ", ", trainLevels,
          "\n")

+       cat("score: ", length(scoreLevels), ", ", scoreLevels,
          "\n\n")

+       newScoreData[,varname] <- factor(newScoreData[,varname],

+                               levels=trainLevels)

+   } #endif

+ }

Warning: new values found in score data, and they will be changed
  to NA!
```

```
GENDER
train:   7,   A   C   F   J   M   U
score:   5,   F   J   M   U
```

```
Warning: new values found in score data, and they will be changed
    to NA!
```

```
STATE
train: 57, AA  AE  AK  AL  AP  AR  AS  AZ  CA  CO  CT  DC  DE  FL  GA  GU  HI
           IA  ID  IL  IN  KS  KY  LA  MA  MD  ME  MI  MN  MO  MS  MT  NC  ND
           NE  NH  NJ  NM  NV  NY  OH  OK  OR  PA  RI  SC  SD  TN  TX  UT  VA
           VI  VT  WA  WI  WV  WY
score: 59, AA  AE  AK  AL  AP  AR  ASv AZ  CA  CO  CT  DC  DE  FL  GA  GU
           HI  IA  ID  IL  IN  KS  KY  LA  MA  MD  ME  MI  MN  MO  MS  MT  NC
           ND  NE  NH  NJ  NM  NV  NY  OH  OK  OR  PA  PR  PW  RI  SC  SD  TN
           TX  UT  VA  VI  VT  WA  WI  WV  WY
```

After checking the new data, we then load the model and check memory usage. We also remove some objects which will no longer be used and do a garbage collection with function gc().

```
> # loading model

> load(paste(treeFileName, ".Rdata", sep=""))

> print(object.size(trainData), units = "Mb")

8 Mb

> print(object.size(scoreData), units = "Mb")

8.1 Mb

> print(object.size(newScoreData), units = "Mb")

8.1 Mb

> print(object.size(myCtree), units = "Mb")

39.7 Mb

> gc()

          used       (Mb)   gc trigger  (Mb)    max used    (Mb)
Ncells    670228     35.8   1073225     57.4    1073225     57.4
Vcells    60433162   461.1  130805779   998.0   130557146   996.1

> memory.size()

[1] 516.73
```

```
> rm(trainNames, scoreNames)

> rm(variableList)

> rm(trainLevels, scoreLevels)

> rm(trainData, scoreData)

> gc()
```

```
          used      (Mb)    gc trigger   (Mb)    max used    (Mb)
Ncells    670071    35.8    1073225      57.4    1073225     57.4
Vcells    58323258  445.0   130805779    998.0   130557146   996.1
```

```
> memory.size()
```

```
[1] 500.23
```

Next, the scoring data are split into multiple subsets and the built tree is applied to each subset to reduce memory consumption. After scoring, the distribution of scores is shown in Figure 14.5.

```
> nScore <- dim(newScoreData)[1]

> (splitSize <- round(nScore/splitNum))

[1] 9637

> myPred <- NULL

> for (i in 1:splitNum) {

+     startPos <- 1 + (i-1)*splitSize

+     if (i==splitNum) {

+         endPos <- nScore

+     }

+     else {

+         endPos <- i * splitSize

+     }

+     print(paste("Predicting:", startPos, "-", endPos))

+     # make prediction

+     tmpPred <- predict(myCtree, newdata = newScoreData
          [startPos:endPos,])
```

```
+      myPred <- c(myPred, tmpPred)

+ }

[1] "Predicting: 1 - 9637"

[1] "Predicting: 9638 - 19274"

[1] "Predicting: 19275 - 28911"

[1] "Predicting: 28912 - 38548"

[1] "Predicting: 38549 - 48185"

[1] "Predicting: 48186 - 57822"

[1] "Predicting: 57823 - 67459"

[1] "Predicting: 67460 - 77096"

[1] "Predicting: 77097 - 86733"

[1] "Predicting: 86734 - 96367"

> # cumulative count and percentage

> length(myPred)

[1] 96367

> rankedLevels <- table(round(myPred, digits=4))

> # put highest rank first by reversing the vector

> rankedLevels <- rankedLevels[length(rankedLevels):1]

> levelNum <- length(rankedLevels)

> cumCnt <- rep(0, levelNum)

> cumCnt[1] <- rankedLevels[1]

> for (i in 2:levelNum) {

+      cumCnt[i] <- cumCnt[i-1] + rankedLevels[i]

+}

> cumPercent <- 100 * cumCnt / nScore

> cumPercent <- round(cumPercent, digits=1)
```

```
> percent <- 100 * rankedLevels / nScore

> percent <- round(percent, digits=1)

> cumRanking <- data.frame(rankedLevels, cumCnt, percent,
  cumPercent)

> names(cumRanking) <- c("Frequency", "CumFrequency",
  "Percentage", "CumPercentage")

> print(cumRanking)
```

	Frequency	CumFrequency	Percentage	CumPercentage
0.8889	9	9	0.0	0.0
0.2124	141	150	0.1	0.2
0.1842	68	218	0.1	0.2
0.135	1342	1560	1.4	1.6
0.1222	1779	3339	1.8	3.5
0.1128	1369	4708	1.4	4.9
0.106	278	4986	0.3	5.2
0.1026	56	5042	0.1	5.2
0.0851	1138	6180	1.2	6.4
0.0764	10603	16783	11.0	17.4
0.0744	800	17583	0.8	18.2
0.0574	4611	22194	4.8	23.0
0.057	8179	30373	8.5	31.5
0.0557	4759	35132	4.9	36.5
0.0535	5558	40690	5.8	42.2
0.0526	1178	41868	1.2	43.4
0.0426	10191	52059	10.6	54.0
0.0378	24757	76816	25.7	79.7
0.0324	13475	90291	14.0	93.7
0.031	1189	91480	1.2	94.9
0.0224	4887	96367	5.1	100.0

```
> write.csv(cumRanking, "cup98-cumulative-ranking.csv",
  row.names=T)

> pdf(paste("cup98-score-distribution.pdf",sep=""))

> plot(rankedLevels, x=names(rankedLevels), type="h",
  xlab="Score", ylab="# of Customers")

> graphics.off()
```

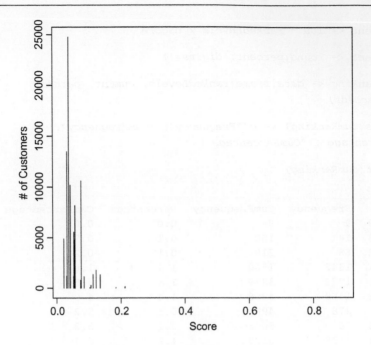

Figure 14.5 Distribution of scores.

Next, we use the predicted scores to rank customers and save the result into a .CSV file.

```
> s1 <- sort(myPred, decreasing=TRUE, method = "quick",
  index.return=T)

> varToOutput <- c("CONTROLN")

> score <- round(myPred[s1$ix], digits=4)

> table(score, useNA="ifany")


score
0.0224 0.031  0.0324 0.0378 0.0426 0.0526 0.0535 0.0557 0.057  0.0574 0.0744
4887   1189   13475  24757  10191  1178   5558   4759   8179   4611   800
0.0764 0.0851 0.1026 0.106  0.1128 0.1222 0.135  0.1842 0.2124 0.8889
10603  1138   56     278    1369   1779   1342   68     141    9


> result <- data.frame(cbind(newScoreData[s1$ix, varToOutput]),
  score)

> names(result) <- c(varToOutput, "score")

> write.csv(result, "cup98-predicted-score.csv", row.names=F)
```

Below is an example for saving result into an EXCEL file.

```
> # output as an EXCEL file

> library(RODBC)

> xlsFile <- odbcConnectExcel("cup98-predicted-score.xls",
  readOnly=F)

> sqlSave(xlsFile, result, rownames=F)

> odbcCloseAll()
```

14.9   Print Rules

This section provides codes for printing rules in the order of their scores. We first present R code for print rules in text, and then demonstrate how to print the rules for running in SAS.

14.9.1   Print Rules in Text

At first, we rewrite the print functions for TerminalNode, SplittingNode, orderedSplit, and nominalSplit, based on the source code in file "Print.R" from package *party*.

```
> # functions for printing rules from ctree

> # based on "Print.R" from package party

> print.TerminalNode <- function(x, rule = NULL, ...) {

+       n.rules ≪- n.rules + 1

+       node.ids ≪- c(node.ids, x$nodeID)

+       n.records ≪- c(n.records, sum(x$weights))

+       scores ≪- c(scores, x$prediction)

+       ruleset ≪- c(ruleset, rule)

+ }

> print.SplittingNode <- function(x, rule = NULL, …) {

+       if (!is.null(rule)) {

+          rule <- paste (rule, "\n")

+       }
```

```
+      rule2 <- print(x$psplit, left = TRUE, rule=rule)

+      print(x$left, rule=rule2)

+      rule3 <- print(x$psplit, left = FALSE, rule=rule)

+      print(x$right, rule=rule3)

+ }

> print.orderedSplit <- function(x, left = TRUE, rule = NULL, …) {

+      if (!is.null (attr (x$splitpoint, "levels"))) {

+        sp <- attr (x$splitpoint, "levels")[x$splitpoint]

+      } else {

+        sp <- x$splitpoint

+      }

+      n.pad <- 20 - nchar (x$variableName)

+      pad <- paste(rep(" ", n.pad), collapse="")

+      if (!is.null(x$toleft)) {

+        left <- as.logical(x$toleft) == left

+      }

+      if (left) {

+        rule2 <- paste(rule, x$variableName, pad, "<= ", sp, sep = "")

+      } else {

+        rule2 <- paste (rule, x$variableName, pad, "> ", sp, sep = "")

+      }

+      rule2

+ }

> print.nominalSplit <- function(x, left = TRUE, rule = NULL, …) {

+      levels <- attr(x$splitpoint, "levels")

+      ### is > 0 for levels available in this node

+      tab <- x$table
```

```
+        if (left) {

+            lev <- levels[as.logical(x$splitpoint) & (tab > 0)]

+        } else {

+            lev <- levels[!as.logical(x$splitpoint) & (tab > 0)]

+        }

+        txt <- paste("'", paste(lev, collapse="', '"), "'", sep="")

+        n.pad <- 20 - nchar(x$variableName)

+        pad <- paste(rep(" ", n.pad), collapse="")

+        rule2 <- paste(rule, x$variableName, pad, txt, sep = "")

+        rule2

+ }
```

After that, by calling function `print(myCtree@tree)`, the information of the tree are extracted and written to five global variables:

- `n.rules`: the number of rules;
- `node.ids`: the IDs of leaf nodes;
- `n.records`: the number of records falling in every leaf node;
- `scores`: the score of every leaf node; and
- `ruleset`: a set of rules corresponding to every leaf node.

```
> library(party) # for ctree

> # loading model

> load(paste(treeFileName, ".Rdata", sep=""))

> # extract rules from tree

> n.rules <- 0

> node.ids <- NULL

> n.records <- NULL

> scores <- NULL

> ruleset <- NULL

> print(myCtree@tree)
```

```
> n.rules
```

```
[1] 21
```

Now all information needed for the rules has been extracted and saved in the above five global variables by calling `print(myCtree@tree)`. The rules are then sorted by score and printed, together with the percentage and cumulative percentage of records covered by the rules. In the code below, function `cumsum()` calculates cumulative sum of a numeric vector. Only the first five rules are printed to save space.

```
> # sort by score descendingly
```

```
> s1 <- sort(scores, decreasing=T, method="quick",
  index.return=T)
```

```
> percentage <- 100 * n.records[s1$ix] / sum(myCtree@weights)
```

```
> cumPercentage <- round(cumsum(percentage), digits=1)
```

```
> percentage <- round(percentage, digits=1)
```

```
> # print all rules
```

```
> for (i in 1:n.rules) {
```

```
+    cat("Rule", i, "\n")
```

```
+    cat("Node:", node.ids[s1$ix[i]])
```

```
+    cat(", score:", scores[s1$ix[i]])
```

```
+    cat(", Percentage: ", percentage[i], "%", sep="")
```

```
+    cat(", Cumulative Percentage: "%",cumPercentage[i], "%",
  sep="")
```

```
+    cat(ruleset[s1$ix[i]], "\n\n")
```

```
+}
```

```
Rule 1
```

```
Node: 27, score: 0.8888889, Percentage: 0%, Cumulaive Percentage:
  0%
```

```
  RFA_2A        'F', 'G'
  PEPSTRFL      'X'
  LASTDATE      <= 9607
  MINRAMNT      > 12.5
  CARDPM12      <= 4
  NGIFTALL      <= 2
```

```
Rule 2

Node: 32, score: 0.2123894, Percentage: 0.1%, Cumulaive
  Percentage: 0.2%


RFA_2A       'F', 'G'
PEPSTRFL     'X'
LASTDATE     > 9607
CARDPM12     > 10


Rule 3

Node: 22, score: 0.1842105, Percentage: 0%, Cumulaive Percentage:
  0.2%


RFA_2A       'F', 'G'
PEPSTRFL     'X'
LASTDATE     <= 9607
MINRAMNT     <= 12.5
RFA_2F       > 1
GENDER       ' ', 'A', 'J', 'M', 'U'
GENDER       'A', 'J'


Rule 4

Node: 14, score: 0.135, Percentage: 1.3%, Cumulaive Percentage:
  1.5%


RFA_2A       'D', 'E'
LASTDATE     > 9606


Rule 5

Node: 8, score: 0.1221591, Percentage: 1.8%, Cumulaive
  Percentage: 3.4%


RFA_2A       'D', 'E'
LASTDATE     <= 9606
RFA_2F       > 2
CARDPM12     <= 4
```

14.9.2   Print Rules for Scoring with SAS

Similar to Section 14.9.1, this section presents R code for printing rules for scoring with SAS in its DATA step. Below are four revised print functions.

```
> # functions for printing rules in SAS statement for scoring
  with a DATA step

> # based on "Print.R" from package party

> print.TerminalNode <- function(x, rule = NULL, …) {

+     rule <- sub(' +', '', rule) # remove leading spaces

+     n.rules ≪ - n.rules + 1

+     node.ids ≪ - c(node.ids, x$nodeID)

+     n.records ≪ - c(n.records, sum(x$weights))

+     scores ≪ - c(scores, x$prediction)

+     ruleset ≪ - c(ruleset, rule)

+ }

> print.SplittingNode <- function(x, rule = NULL, …) {

+     if (!is.null (rule)) {

+        rule <- paste(rule, "\n and")

+     }#endif

+     rule2 <- print(x$psplit, left = TRUE, rule=rule)

+     print(x$left, rule=rule2)

+     rule3 <- print(x$psplit, left = FALSE, rule=rule)

+     print(x$right, rule=rule3)

+}

> print.orderedSplit <- function(x, left = TRUE, rule = NULL, …) {

+     if (!is.null (attr (x$splitpoint, "levels"))) {

+        sp <- attr (x$splitpoint, "levels")[x$splitpoint]

+     } else {

+         sp <- x$splitpoint

+     }

+     if (!is.null(x$toleft)) left <- as.logical(x$toleft) ==
  left
```

```
+      if (left) {

+        rule2 <- paste(rule, " ", x$variableName, " <= ", sp, sep =
  "")

+      } else {

+          rule2 <- paste(rule, " ", x$variableName, " > ", sp, sep
  = "")

+      }

+      rule2

+}

> print.nominalSplit <- function(x, left = TRUE, rule = NULL, ...) {

+      levels <- attr(x$splitpoint, "levels")

+      ### is > 0 for levels available in this node

+      tab <- x$table

+      if (left) {

+        lev <- levels[as.logical(x$splitpoint) & (tab > 0)]

+      } else {

+        lev <- levels[!as.logical(x$splitpoint) & (tab > 0)]

+      }

+      txt <- paste("'", paste(lev, collapse="', '"), "'", sep="")

+      rule2 <- paste(rule, " ", x$variableName, " in (", txt,
  ")", sep = "")

+      rule2

+}
```

Again, by calling function `print(myCtree@tree)`, the information of the tree are extracted and written to five global variables: `n.rules`, `node.ids`, `n.records`, `scores`, and `ruleset`.

```
> library(party) # for ctree

> # loading model

> load(paste(treeFileName, ".Rdata", sep=""))
```

```
> n.rules <- 0

> node.ids <- NULL

> n.records <- NULL

> scores <- NULL

> ruleset <- NULL

> print(myCtree@tree)

> n.rules

[1] 21
```

The rules are then sorted by score and printed. Only the first five rules are shown below to save space. The printed rules can be copied and pasted in a SAS DATA step for scoring new data.

```
> # sort by score descendingly

> s1 <- sort(scores, decreasing=TRUE, method = "quick",
  index.return=TRUE)

> percentage <- 100 * n.records[s1$ix]/ sum (myCtree@weights)

> cumPercentage <- round(cumsum(percentage), digits=1)

> percentage <- round(percentage, digits=1)

> # print all rules

> for (i in 1:n.rules) {

+     cat("/* Rule", i, "\n")

+     cat("   Node:", node.ids[s1$ix[i]])

+     cat(", score:", scores[s1$ix[i]])

+     cat(", Percentage: ", percentage[i], "%", sep=" ")

+     cat(", Cumulative Percentage: ",cumPercentage[i], "% \n*/\n",
  sep=" ")

+     if(i == 1){

+       cat("IF \n   ")

+     } else {

+       cat("ELSE IF \n   ")
```

```
+        }

+        cat(ruleset[s1$ix[i]], "\n")

+        cat("THEN\n  score = ", scores[s1$ix[i]], ";\n\n", sep="")

+        }
/* Rule 1

     Node: 27, score: 0.8888889, Percentage: 0%, Cumalative
  Percentage: 0%

*/

IF

    RFA_2A in ('F', 'G')

    and PEPSTRFL in ('X')

    and LASTDATE <= 9607

    and MINRAMNT > 12.5

    and CARDPM12 <= 4

    and NGIFTALL <= 2

THEN
  score = 0.8888889;

/* Rule 2

     Node: 32, score: 0.2123894, Percentage: 0.1%, cumalative
  percentage: 0.2%

*/

ELSE IF

    RFA_2A in ('F', 'G')

    and PEPSTRFL in ('X')

    and LASTDATE > 9607

    and CARDPM12 > 10

THEN

    score = 0.2123894;

/* Rule 3
```

```
    Node: 22, score: 0.1842105, Percentage: 0%, cumalative
  percentage: 0.2%

*/

ELSE IF

    RFA_2A in ('F', 'G')

    and PEPSTRFL in ('X')

    and LASTDATE <= 9607

    and MINRAMNT <= 12.5

    and RFA_2F > 1

    and GENDER in (' ', 'A', 'J', 'M', 'U')

    and GENDER in ('A', 'J')

THEN

    score = 0.1842105;

/* Rule 4

Node: 14, score: 0.135, Percentage: 1.3%, Cumalative Percentage:
  1.5%

*/

  ELSE IF

    RFA_2A in ('D', 'E')

    and LASTDATE > 9606

THEN

score = 0.135;

/* Rule 5

    Node: 8, score: 0.1221591, Percentage: 1.8%, Cumulative
  Percentage: 3.4%

*/

ELSE IF

    RFA_2A in ('D', 'E')

    and LASTDATE <= 9606
```

```
and RFA_2F > 2

and CARDPM12 <= 4
```

THEN

```
score = 0.1221591;
```

14.10   Conclusions and Discussion

This chapter presents a case study on predictive modeling of big data with limited memory. By building a number of trees on sampled data, useful variables are found and collected. These variables are then used to build a final model. This methodology enables modeling of big data with limited amount of memory.

Another way is sampling variables, instead of sampling records. The variables are sampled each time to build a model. After 10 or 20 models are built, useful variables can be collected from those models and then used to build a final model. This method is similar to the idea of random forest where each tree in the forest is built with a random subset of variables. However, it will use much less memory than random forest.

15 Online Resources

This chapter presents links to online resources on R and data mining, includes books, documents, tutorials, and slides. A list of links is also available at http://www.rdata mining.com/resources/onlinedocs.

15.1 R Reference Cards

- *R Reference Card*, by Tom Short
  http://cran.r-project.org/doc/contrib/Short-refcard.pdf

- *R Reference Card for Data Mining*, by Yanchang Zhao
  http://www.rdatamining.com/docs

- *R Reference Card*, by Jonathan Baron
  http://cran.r-project.org/doc/contrib/refcard.pdf

- *R Functions for Regression Analysis*, by Vito Ricci
  http://cran.r-project.org/doc/contrib/Ricci-refcard-regression.pdf

- *R Functions for Time Series Analysis*, by Vito Ricci
  http://cran.r-project.org/doc/contrib/Ricci-refcard-ts.pdf

15.2 R

- *Quick-R*
  http://www.statmethods.net/

- *R Tips*: lots of tips for R programming
  http://pj.freefaculty.org/R/Rtips.html

- *R Tutorial*
  http://www.cyclismo.org/tutorial/R/index.html

- *The R Manuals*, including *an Introduction to R, R Language Definition, R Data Import/Export*, and other R manuals
  http://cran.r-project.org/manuals.html

- *R You Ready?*

R and Data Mining. DOI: http://dx.doi.org/10.1016/B978-0-12-396963-7.00015-5
© 2013 Yanchang Zhao. Published by Elsevier Inc. All rights reserved.

```
http://pj.freefaculty.org/R/RUReady.pdf
```

- *R for Beginners*
  ```
  http://cran.r-project.org/doc/contrib/Paradis-rdebuts_en.pdf
  ```

- *Econometrics in R*
  ```
  http://cran.r-project.org/doc/contrib/Farnsworth-EconometricsInR.
  pdf
  ```

- *Using R for Data Analysis and Graphics—Introduction, Examples, and Commentary*
  ```
  http://www.cran.r-project.org/doc/contrib/usingR.pdf
  ```

- Lots of R Contributed Documents, including non-English ones
  ```
  http://cran.r-project.org/other-docs.html
  ```

- *The R Journal*
  ```
  http://journal.r-project.org/current.html
  ```

- *Learn R Toolkit*
  ```
  http://processtrends.com/Learn_R_Toolkit.htm
  ```

- *Resources to help you learn and use R* at UCLA
  ```
  http://www.ats.ucla.edu/stat/r/
  ```

- *R Tutorial—An R Introduction to Statistics*
  ```
  http://www.r-tutor.com/
  ```

- *Cookbook for R*
  ```
  http://wiki.stdout.org/rcookbook/
  ```

- *Slides for a couple of R short courses*
  ```
  http://courses.had.co.nz/
  ```

- *Tips on memory in R*
  ```
  http://www.matthewckeller.com/html/memory.html
  ```

15.3 Data Mining

- *Introduction to Data Mining*, by Pang-Ning Tan, Michael Steinbach, and Vipin Kumar
  ```
  http://www-users.cs.umn.edu/%7Ekumar/dmbook
  ```

- *Tutorial on Data Mining Algorithms* by Ian Witten
  ```
  http://www.cs.waikato.ac.nz/~ihw/DataMiningTalk/
  ```

- *Mining of Massive Datasets*, by Anand Rajaraman and Jeff Ullman
  http://infolab.stanford.edu/%7Eullman/mmds.html

- *Lecture notes of data mining course*, by Cosma Shalizi at CMU
  http://www.stat.cmu.edu/%7Ecshalizi/350/

- *Introduction to Information Retrieval*, by Christopher D. Manning, Prabhakar Raghavan, and Hinrich Schütze at Stanford University
  http://nlp.stanford.edu/IR-book/

- *Statistical Data Mining Tutorials*, by Andrew Moore
  http://www.autonlab.org/tutorials/

- *Tutorial on Spatial and Spatio-Temporal Data Mining*
  http://www.inf.ufsc.br/%7Evania/tutorial_icdm.html

- *Tutorial on Discovering Multiple Clustering Solutions*
  http://dme.rwth-aachen.de/en/DMCS

- *Time-Critical Decision Making for Business Administration*
  http://home.ubalt.edu/ntsbarsh/stat-data/Forecast.htm

- A paper on *Open-Source Tools for Data Mining*
  http://eprints.fri.uni-lj.si/893/1/2008-OpenSourceDataMining.pdf

- *An overview of data mining tools*
  http://onlinelibrary.wiley.com/doi/10.1002/widm.24/pdf

- *Textbook on Introduction to social network methods*
  http://www.faculty.ucr.edu/~hanneman/nettext/

- *Information Diffusion In Social Networks: Observing and Influencing Societal Interests*, a tutorial at VLDB'11
  http://www.cs.ucsb.edu/~cbudak/vldb_tutorial.pdf

- *Tools for large graph mining: structure and diffusion*, a tutorial at WWW2008
  http://cs.stanford.edu/people/jure/talks/www08tutorial/

- *Graph Mining: Laws, Generators, and Tools*
  http://www.stanford.edu/group/mmds/slides2008/faloutsos.pdf

- *A tutorial on outlier detection techniques* at ACM SIGKDD'10
  http://www.dbs.ifi.lmu.de/~zimek/publications/KDD2010/kdd10-outlier-tutorial.pdf

- *A Taste of Sentiment Analysis* - 105-page slides in PDF format
  http://statmath.wu.ac.at/research/talks/resources/sentimentanalysis.pdf

15.4   Data Mining with R

- *Data Mining with R—Learning by Case Studies*
  http://www.liaad.up.pt/~ltorgo/DataMiningWithR/

- *Data Mining Algorithms In R*
  http://en.wikibooks.org/wiki/Data_Mining_Algorithms_In_R

- *Statistics with R*
  http://zoonek2.free.fr/UNIX/48_R/all.html

- *Data Mining Desktop Survival Guide*
  http://www.togaware.com/datamining/survivor/

15.5   Classification/Prediction with R

- *An Introduction to Recursive Partitioning Using the RPART Routines*
  http://www.mayo.edu/hsr/techrpt/61.pdf

- *Visualizing classifier performance with package ROCR*
  http://rocr.bioinf.mpi-sb.mpg.de/ROCR_Talk_Tobias_Sing.ppt

15.6   Time Series Analysis with R

- *An R Time Series Tutorial*
  http://www.stat.pitt.edu/stoffer/tsa2/R_time_series_quick_fix.htm

- *Time Series Analysis with R*
  http://www.statoek.wiso.uni-goettingen.de/veranstaltungen/zeitr
  eihen/sommer03/ts_r_intro.pdf

- *Using R (with applications in Time Series Analysis)*
  http://people.bath.ac.uk/masgs/time%20series/TimeSeriesR2004.pdf

- *CRAN Task View: Time Series Analysis*
  http://cran.r-project.org/web/views/TimeSeries.html

15.7   Association Rule Mining with R

- *Introduction to arules: A computational environment for mining association rules and frequent item sets*
  http://cran.csiro.au/web/packages/arules/vignettes/arules.pdf

- *Visualizing Association Rules: Introduction to arulesViz*
  http://cran.csiro.au/web/packages/arulesViz/vignettes/arulesViz.
  pdf

- *Association Rule Algorithms In R*
  http://en.wikibooks.org/wiki/Data_Mining_Algorithms_In_R/
  Frequent_Pattern_Mining

15.8   Spatial Data Analysis with R

- *Applied Spatio-temporal Data Analysis with FOSS: R+OSGeo*
  http://www.geostat-course.org/GeoSciences_AU_2011

- *Spatial Regression Analysis in R—A Workbook*
  http://geodacenter.asu.edu/system/files/rex1.pdf

15.9   Text Mining with R

- *Text Mining Infrastructure in R*
  http://www.jstatsoft.org/v25/i05

- *Introduction to the tm Package Text Mining in R*
  http://cran.r-project.org/web/packages/tm/vignettes/tm.pdf

- *Text Mining Handbook* with R code examples
  http://www.casact.org/pubs/forum/10spforum/Francis_Flynn.pdf

- *Distributed Text Mining in R*
  http://epub.wu.ac.at/3034/

15.10   Social Network Analysis with R

- *R for networks: a short tutorial*
  http://sites.stat.psu.edu/~dhunter/Rnetworks/

- *Social Network Analysis in R*
  http://files.meetup.com/1406240/sna_in_R.pdf

- *A detailed introduction to Social Network Analysis with package sna*
  http://www.jstatsoft.org/v24/i06/paper

- *A statnet Tutorial*
  http://www.jstatsoft.org/v24/i09/paper

- *Slides on Social network analysis with R*
  http://user2010.org/slides/Zhang.pdf

- *Tutorials on using statnet for network analysis*
  http://csde.washington.edu/statnet/resources.shtml

15.11   Data Cleansing and Transformation with R

- *Tidy Data and Tidy Tools*
  http://vita.had.co.nz/papers/tidy-data-pres.pdf

- *The data.table package in R*
  http://files.meetup.com/1677477/R_Group_June_2011.pdf

15.12   Big Data and Parallel Computing with R

- *State of the Art in Parallel Computing with R*
  http://www.jstatsoft.org/v31/i01/paper

- *Taking R to the Limit, Part I—Parallelization in R*
  http://www.bytemining.com/2010/07/taking-r-to-the-limit-part-i-parallelization-in-r/

- *Taking R to the Limit, Part II—Large Datasets in R*
  http://www.bytemining.com/2010/08/taking-r-to-the-limit-part-ii-large-datasets-in-r/

- *Tutorial on MapReduce programming in R with package rmr*
  https://github.com/RevolutionAnalytics/RHadoop/wiki/Tutorial

- *Distributed Data Analysis with Hadoop and R*
  http://www.infoq.com/presentations/Distributed-Data-Analysis-with-Hadoop-and-R

- *Massive data, shared and distributed memory, and concurrent programming: big-memory and foreach*
  http://sites.google.com/site/bigmemoryorg/research/documentation/bigmemorypresentation.pdf

- *High Performance Computing with R*
  http://igmcs.utk.edu/sites/igmcs/files/Patel-High-Performance-Computing-with-R-2011-10-20.pdf

- *R with High Performance Computing: Parallel processing and large memory*
  `http://files.meetup.com/1781511/HighPerformanceComputingR-`
  `Szczepanski.pdf`

- *Parallel Computing in R*
  `http://blog.revolutionanalytics.com/downloads/BioC2009%20ParallelR.`
  `pdf`

- *Parallel Computing with R using snow and snowfall*
  `http://www.ics.uci.edu/~vqnguyen/talks/ParallelComputingSeminaR.`
  `pdf`

- *Interacting with Data using the filehash Package for R*
  `http://cran.r-project.org/web/packages/filehash/vignettes/fileh`
  `ash.pdf`

- *Tutorial: Parallel computing using R package snowfall*
  `http://www.imbi.uni-freiburg.de/parallel/docs/Reisensburg2009_`
  `TutParallelComputing_Knaus_Porzelius.pdf`

- *Easier Parallel Computing in R with snowfall and sfCluster*
  `http://journal.r-project.org/2009-1/RJournal_2009-1_Knaus+et+al.`
  `pdf`

R Reference Card for Data Mining

by Yanchang Zhao, yanchang@rdatamining.com, August 11, 2012
The latest version is available at http://www.RDataMining.com. Click the link also for document *R and Data Mining: Examples and Case Studies*. The package names are in parentheses.

Association Rules & Frequent Itemsets

APRIORI Algorithm
a level-wise, breadth-first algorithm which counts transactions to find frequent itemsets
apriori() mine associations with APRIORI algorithm (*arules*)

ECLAT Algorithm
employs equivalence classes, depth-first search and set intersection instead of counting
eclat() mine frequent itemsets with the Eclat algorithm (*arules*)

Packages
arules mine frequent itemsets, maximal frequent itemsets, closed frequent itemsets and association rules. It includes two algorithms, Apriori and Eclat.
arulesViz visualizing association rules

Sequential Patterns

Functions
cspade() mining frequent sequential patterns with the cSPADE algorithm (*arulesSequences*)
seqefsub() searching for frequent subsequences (*TraMineR*)

Packages
arulesSequences add-on for *arules* to handle and mine frequent sequences
TraMineR mining, describing and visualizing sequences of states or events

Classification & Prediction

Decision Trees
ctree() conditional inference trees, recursive partitioning for continuous, censored, ordered, nominal and multivariate response variables in a conditional inference framework (*party*)
rpart() recursive partitioning and regression trees (*rpart*)
mob() model-based recursive partitioning, yielding a tree with fitted models associated with each terminal node (*party*)

Random Forest
cforest() random forest and bagging ensemble (*party*)
randomForest() random forest (*randomForest*)
varimp() variable importance (*party*)
importance() variable importance (*randomForest*)

Neural Networks
nnet() fit single-hidden-layer neural network (*nnet*)

Support Vector Machine (SVM)
svm() train a support vector machine for regression, classification or density-estimation (*e1071*)
ksvm() support vector machines (*kernlab*)

Performance Evaluation
performance() provide various measures for evaluating performance of prediction and classification models (*ROCR*)
roc() build a ROC curve (*pROC*)
auc() compute the area under the ROC curve (*pROC*)
ROC() draw a ROC curve (*DiagnosisMed*)
PRcurve() precision-recall curves (*DMwR*)
CRchart() cumulative recall charts (*DMwR*)

Packages
rpart recursive partitioning and regression trees
party recursive partitioning
randomForest classification and regression based on a forest of trees using random inputs
rpartOrdinal ordinal classification trees, deriving a classification tree when the response to be predicted is ordinal
rpart.plot plots rpart models with an enhanced version of plot.rpart in the rpart package
ROCR visualize the performance of scoring classifiers
pROC display and analyze ROC curves

Regression

Functions
lm() linear regression
glm() generalized linear regression
nls() non-linear regression
predict() predict with models
residuals() residuals, the difference between observed values and fitted values
gls() fit a linear model using generalized least squares (*nlme*)
gnls() fit a nonlinear model using generalized least squares (*nlme*)

Packages
nlme linear and nonlinear mixed effects models

Clustering

Partitioning based Clustering
partition the data into k groups first and then try to improve the quality of clustering by moving objects from one group to another
kmeans() perform k-means clustering on a data matrix
kmeansCBI() interface function for clustering methods (*fpc*)
kmeansruns() call kmeans for the k-means clustering method and includes estimation of the number of clusters and finding an optimal solution from several starting points (*fpc*)
pam() the Partitioning Around Medoids (PAM) clustering method (*cluster*)
pamk() the Partitioning Around Medoids (PAM) clustering method with estimation of number of clusters (*fpc*)

cluster.optimal() search for the optimal k-clustering of the dataset (*bayesclust*)
clara() Clustering Large Applications (*cluster*)
fanny(x,k,...) compute a fuzzy clustering of the data into k clusters (*cluster*)
kcca() k-centroids clustering (*flexclust*)
ccfkms() clustering with Conjugate Convex Functions
apcluster() affinity propagation clustering for a given similarity matrix (*apcluster*)
apclusterK() affinity propagation clustering to get K clusters (*apcluster*)
cclust() Convex Clustering, incl. k-means and two other clustering algorithms (*cclust*)
KMeansSparseCluster() sparse k-means clustering (*sparcl*)
tclust(x,k,alpha,...) trimmed k-means with which a proportion alpha of observations may be trimmed (*tclust*)

Hierarchical Clustering
a hierarchical decomposition of data in either bottom-up (agglomerative) or top-down (divisive) way
hclust(d, method, ...) hierarchical cluster analysis on a set of dissimilarities d using the method for agglomeration
pvclust() hierarchical clustering with p-values via multi-scale bootstrap resampling (*pvclust*)
agnes() agglomerative hierarchical clustering (*cluster*)
diana() divisive hierarchical clustering (*cluster*)
mona() divisive hierarchical clustering of a dataset with binary variables only (*cluster*)
rockCluster() cluster a data matrix using the Rock algorithm (*cba*)
proximus() cluster the rows of a logical matrix using the Proximus algorithm (*cba*)
isopam() Isopam clustering algorithm (*isopam*)
LLAhclust() hierarchical clustering based on likelihood linkage analysis (*LLAhclust*)
flashClust() optimal hierarchical clustering (*flashClust*)
fastcluster() fast hierarchical clustering (*fastcluster*)
cutreeDynamic(), cutreeHybrid() detection of clusters in hierarchical clustering dendrograms (*dynamicTreeCut*)
HierarchicalSparseCluster() hierarchical sparse clustering (*sparcl*)

Model based Clustering
Mclust() model-based clustering (*mclust*)
HDDC() a model-based method for high dimensional data clustering (*HDclassif*)
fixmahal() Mahalanobis Fixed Point Clustering (*fpc*)
fixreg() Regression Fixed Point Clustering (*fpc*)
mergenormals() clustering by merging Gaussian mixture components (*fpc*)

Density based Clustering
generate clusters by connecting dense regions
dbscan(data,eps,MinPts,...) generate a density based clustering of arbitrary shapes, with neighborhood radius set as eps and density thresh-

old as MinPts (*fpc*)

pdfCluster() clustering via kernel density estimation (*pdfCluster*)

Other Clustering Techniques

mixer() random graph clustering (*mixer*)

nncluster() fast clustering with restarted minimum spanning tree (*nnclust*)

orclus() ORCLUS subspace clustering (*orclus*)

Plotting Clustering Solutions

plotcluster() visualisation of a clustering or grouping in data (*fpc*)

bannerplot() a horizontal barplot visualizing a hierarchical clustering (*cluster*)

Cluster Validation

silhouette() compute or extract silhouette information (*cluster*)

cluster.stats() compute several cluster validity statistics from a clustering and a dissimilarity matrix (*fpc*)

clValid() calculate validation measures for a given set of clustering algorithms and number of clusters (*clValid*)

clustIndex() calculate the values of several clustering indexes, which can be independently used to determine the number of clusters existing in a data set

NbClust() provide 30 indices for cluster validation and determining the number of clusters (*NbClust*)

Packages

cluster cluster analysis

fpc various methods for clustering and cluster validation

mclust model-based clustering and normal mixture modeling

birch clustering very large datasets using the BIRCH algorithm

pvclust hierarchical clustering with p-values

apcluster Affinity Propagation Clustering

cclust Convex Clustering methods, including k-means algorithm, On-line Update algorithm and Neural Gas algorithm and calculation of indexes for finding the number of clusters in a data set

cba Clustering for Business Analytics, including clustering techniques such as Proximus and Rock

bclust Bayesian clustering using spike-and-slab hierarchical model, suitable for clustering high-dimensional data

biclust algorithms to find bi-clusters in two-dimensional data

clue cluster ensembles

clues clustering method based on local shrinking

clValid validation of clustering results

clv cluster validation techniques, contains popular internal and external cluster validation methods for outputs produced by package *cluster*

bayesclust tests/searches for significant clusters in genetic data

clustvarsel variable selection for model-based clustering

clustsig significant cluster analysis, tests to see which (if any) clusters are statistically different

clusterfly explore clustering interactively

clusterSim search for optimal clustering procedure for a data set

clusterGeneration random cluster generation

clusterCons calculate the consensus clustering result from re-sampled clustering experiments with the option of using multiple clustering algorithms and parameter

gcExplorer graphical cluster explorer

hybridHclust hybrid hierarchical clustering via mutual clusters

Modalclust hierarchical Modal Clustering

iCluster integrative clustering of multiple genomic data types

EMCC evolutionary Monte Carlo (EMC) methods for clustering

rEMM extensible Markov Model (EMM) for data stream clustering

Outlier Detection

Functions

boxplot.stats()$out list data points lying beyond the extremes of the whiskers

lofactor() calculate local outlier factors using the LOF algorithm (*DMwR* or *dprep*)

lof() a parallel implementation of the LOF algorithm (*Rlof*)

Packages

extremevalues detect extreme values in one-dimensional data

mvoutlier multivariate outlier detection based on robust methods

outliers some tests commonly used for identifying outliers

Rlof a parallel implementation of the LOF algorithm

Time Series Analysis

Construction & Plot

ts() create time-series objects (*stats*)

plot.ts() plot time-series objects (*stats*)

smoothts() time series smoothing (*ast*)

sfilter() remove seasonal fluctuation using moving average (*ast*)

Decomposition

decomp() time series decomposition by square-root filter (*timsac*)

decompose() classical seasonal decomposition by moving averages (*stats*)

stl() seasonal decomposition of time series by loess (*stats*)

tsr() time series decomposition (*ast*)

ardec() time series autoregressive decomposition (*ArDec*)

Forecasting

arima() fit an ARIMA model to a univariate time series (*stats*)

predict.Arima() ARIMA forecast from models fitted by arima (*stats*)

auto.arima() fit best ARIMA model to univariate time series (*forecast*)

Packages

timsac time series analysis and control program

ast time series analysis

ArDec time series autoregressive-based decomposition

ares a toolbox for time series analyses using generalized additive models

dse tools for multivariate, linear, time-invariant, time series models

forecast displaying and analysing univariate time series forecasts

Text Mining

Functions

Corpus() build a corpus, which is a collection of text documents (*tm*)

tm_map() transform text documents, e.g., stemming, stopword removal (*tm*)

tm_filter() filtering out documents (*tm*)

TermDocumentMatrix(), DocumentTermMatrix() construct a term-document matrix or a document-term matrix (*tm*)

Dictionary() construct a dictionary from a character vector or a term-document matrix (*tm*)

findAssocs() find associations in a term-document matrix (*tm*)

findFreqTerms() find frequent terms in a term-document matrix (*tm*)

stemDocument() stem words in a text document (*tm*)

stemCompletion() complete stemmed words (*tm*)

termFreq() generate a term frequency vector from a text document (*tm*)

stopwords(language) return stopwords in different languages (*tm*)

removeNumbers(), removePunctuation(), removeWords() remove numbers, punctuation marks, or a set of words from a text document (*tm*)

removeSparseTerms() remove sparse terms from a term-document matrix (*tm*)

textcat() n-gram based text categorization (*textcat*)

SnowballStemmer() Snowball word stemmers (*Snowball*)

LDA() fit a LDA (latent Dirichlet allocation) model (*topicmodels*)

CTM() fit a CTM (correlated topics model) model (*topicmodels*)

terms() extract the most likely terms for each topic (*topicmodels*)

topics() extract the most likely topics for each document (*topicmodels*)

Packages

tm a framework for text mining applications

lda fit topic models with LDA

topicmodels fit topic models with LDA and CTM

tm.plugin.dc a plug-in for package *tm* to support distributed text mining

tm.plugin.mail a plug-in for package *tm* to handle mail

RcmdrPlugin.TextMining GUI for demonstration of text mining concepts and *tm* package

textir a suite of tools for inference about text documents and associated sentiment

tau utilities for text analysis

textcat n-gram based text categorization

YjdnJlp Japanese text analysis by Yahoo! Japan Developer Network

Social Network Analysis and Graph Mining

Functions

graph(), graph.edgelist(), graph.adjacency(), graph.incidence() create graph objects respectively from edges, an edge list, an adjacency matrix and an incidence matrix (*igraph*)

plot(), tkplot() static and interactive plotting of graphs (*igraph*)

gplot(), gplot3d() plot graphs (*sna*)

V(), E() vertex/edge sequence of igraph (*igraph*)

are.connected() check whether two nodes are connected (*igraph*)

degree(), betweenness(), closeness() various centrality scores (*igraph, sna*)

add.edges(), add.vertices(), delete.edges(), delete.vertices() add and delete edges and vertices (*igraph*)

neighborhood() neighborhood of graph vertices (*igraph, sna*)

get.adjlist() adjacency lists for edges or vertices (*igraph*)

Packages

lattice a powerful high-level data visualization system, with an emphasis on multivariate data

ggplot2 an implementation of the Grammar of Graphics

vcd visualizing categorical data

dendro visualization of multivariate, functions, sets, and data

iplots interactive graphics

googleVis an interface between R and the Google Visualisation API to create interactive charts

Data Manipulation

Functions

`transform()` transform a data frame

`scale()` scaling and centering of matrix-like objects

`t()` matrix transpose

`aperm()` array transpose

`sample()` sampling

`table()`, `tabulate()`, `xtabs()` cross tabulation (*stats*)

`stack()`, `unstack()` stacking vectors

`reshape()` reshape a data frame between "wide" format and "long" format (*stats*)

`merge()` merge two data frames

`aggregate()` compute summary statistics of data subsets (*stats*)

`by()` apply a function to a data frame split by factors

`melt()`, `cast()` melt and then cast data into the reshaped or aggregated form you want (*reshape*)

`na.fail`, `na.omit`, `na.exclude`, `na.pass` handle missing values

Packages

reshape flexibly restructure and aggregate data

data.table extension of data.frame for fast indexing, ordered joins, assignment, and grouping and list columns

gdata various tools for data manipulation

Data Access

Functions

`save()`, `load()` save and load R data objects

`read.csv()`, `write.csv()` import from and export to .CSV files

`read.table()`, `write.table()`, `scan()`, `write()` read and write data

`write.matrix()` write a matrix or data frame (*MASS*)

`sqlQuery()` submit an SQL-query to an ODBC database (*RODBC*)

`sqlFetch()` read a table from an ODBC database (*RODBC*)

`odbcConnect()`, `odbcClose()`, `odbcCloseAll()` open/close connections to ODBC databases (*RODBC*)

`dbSendQuery` execute an SQL statement on a given database connection (*DBI*)

`dbConnect()`, `dbDisconnect()` create/close a connection to a DBMS (*DBI*)

Packages

RODBC ODBC database access

Statistical Test

`t.test()` student's t-test (*stats*)

`prop.test()` test of equal or given proportions (*stats*)

`binom.test()` exact binomial test (*stats*)

Mixed Effects Models

`lme()` fit a linear mixed-effects model (*nlme*)

`nlme()` fit a nonlinear mixed-effects model (*nlme*)

Principal Components and Factor Analysis

`princomp()` principal components analysis (*stats*)

`prcomp()` principal components analysis (*stats*)

Other Functions

`var()`, `cov()`, `cor()` variance, covariance, and correlation (*stats*)

`density()` compute kernel density estimates (*stats*)

Packages

nlme linear and nonlinear mixed effects models

Graphics

Functions

`plot()` generic function for plotting (*graphics*)

`barplot()`, `pie()`, `hist()` bar chart, pie chart and histogram (*graphics*)

`boxplot()` box-and-whisker plot (*graphics*)

`stripchart()` one dimensional scatter plot (*graphics*)

`dotchart()` Cleveland dot plot (*graphics*)

`qqnorm()`, `qqplot()`, `qqline()` QQ (quantile-quantile) plot (*stats*)

`coplot()` conditioning plot (*graphics*)

`splom()` conditional scatter plot matrices (*lattice*)

`pairs()` a matrix of scatterplots (*graphics*)

`cpairs()` enhanced scatterplot matrix (*gclus*)

`parcoord()` parallel coordinate plot (*MASS*)

`cparcoord()` enhanced parallel coordinate plot (*gclus*)

`paracoor()` parallel coordinates plot (*denpro*)

`parallelplot()` parallel coordinates plot (*lattice*)

`densityplot()` kernel density plot (*lattice*)

`contour()`, `filled.contour()` contour plot (*graphics*)

`levelplot()`, `contourplot()` level plots and contour plots (*lattice*)

`sunflowerplot()` a sunflower scatter plot (*graphics*)

`assocplot()` association plot (*graphics*)

`mosaicplot()` mosaic plot (*graphics*)

`matplot()` plot the columns of one matrix against the columns of another matrix (*graphics*)

`fourfoldplot()` a fourfold display of a 2 × 2 × k contingency table (*graphics*)

`persp()` perspective plots of surfaces over the x-y plane (*graphics*)

`cloud()`, `wireframe()` 3d scatter plots and surfaces (*lattice*)

`interaction.plot()` two-way interaction plot (*stats*)

`iplot()`, `ihist()`, `ibar()`, `ipcp()` interactive scatter plot, histogram, bar plot, and parallel coordinates plot (*iplots*)

`pdf()`, `postscript()`, `win.metafile()`, `jpeg()`, `bmp()`, `png()`, `tiff()` save graphs into files of various formats

---

`nei()`, `adj()`, `from()`, `to()` vertex/edge sequence indexing (*igraph*)

`cliques()` find cliques, ie. complete subgraphs (*graph*)

`clusters()` maximal connected components of a graph (*igraph*)

`%->%`, `%<-%`, `%--%` edge sequence indexing (*igraph*)

`get.edgelist()` return an edge list in a two-column matrix (*igraph*)

`read.graph()`, `write.graph()` read and writ graphs from and to files (*igraph*)

Packages

sna social network analysis

igraph network analysis and visualization

statnet a set of tools for the representation, visualization, analysis and simulation of network data

egonet ego-centric measures in social network analysis

snort social network-analysis on relational tables

network tools to create and modify network objects

bipartite visualising bipartite networks and calculating some (ecological) indices

blockmodeling generalized and classical blockmodeling of valued networks

diagram visualising simple graphs (networks), plotting flow diagrams

NetCluster clustering for networks

NetData network data for McFarland's SNA R labs

NetIndices estimating network indices, including trophic structure of cfoodwebs in R

NetworkAnalysis statistical inference on populations of weighted or unweighted networks

tnet analysis of weighted, two-mode, and longitudinal networks

triads triad census for networks

Spatial Data Analysis

Functions

`geocode()` geocodes a location using Google Maps (*ggmap*)

`cmap()` quick map plot (*ggmap*)

`get.map()` queries the Google Maps, OpenStreetMap, or Stamen Maps server for a map at a certain location (*ggmap*)

Packages

plotGoogleMaps plot spatial data as HTML map mushup over Google Maps

plotKML visualization of spatial and spatio-temporal objects in Google Earth

ggmap Spatial visualization with Google Maps and OpenStreetMap

clustTool GUI for clustering data with spatial information

SGCS Spatial Graph based Clustering Summaries for spatial point patterns

spdep spatial dependence: weighting schemes, statistics and models

Statistics

Summarization

`summary()` summarize data

`describe()` concise statistical description of data (*Hmisc*)

`boxplot.stats()` box plot statistics

Analysis of Variance

`aov()` fit an analysis of variance model (*stats*)

`anova()` compute analysis of variance (or deviance) tables for one or more fitted model objects (*stats*)

DBI a database interface (DBI) between R and relational DBMS

RMySQL interface to the MySQL database

RJDBC access to databases through the JDBC interface

RSQLite SQLite interface for R

ROracle Oracle database interface (DBI) driver

RpgSQL DBI/RJDBC interface to PostgreSQL database

RODM interface to Oracle Data Mining

xlsReadWrite read and write Excel files

WriteXLS create Excel 2003 (XLS) files from data frames

Big Data

Functions

`big.matrix()` create a standard big.matrix, which is constrained to available RAM (*bigmemory*)

`filebacked.big.matrix()` create a file-backed big.matrix, which may exceed available RAM by using hard drive space (*bigmemory*)

`which()` expanded "which"-like functionality (*bigmemory*)

Packages

ff memory-efficient storage of large data on disk and fast access functions

filehash a simple key-value database for handling large data

g.data create and maintain delayed-data packages

BufferedMatrix a matrix data storage object held in temporary files

biglm regression for data too large to fit in memory

bigmemory manage massive matrices with shared memory and memory-mapped files

biganalytics extend the *bigmemory* package with various analytics

bigtabulate table-, tapply-, and split-like functionality for matrix and big.matrix objects

Parallel Computing

Functions

`foreach(...)  %dopar%` looping in parallel (*foreach*)

`registerDoSEQ()`, `registerDoSNOW()`, `registerDoMC()` register respectively the sequential, SNOW and multicore parallel backend with the *foreach* package (*foreach*, *doSNOW*, *doMC*)

`sfInit()`, `sfStop()` initialize and stop the cluster (*snowfall*)

`sfLapply()`, `sfSapply()`, `sfApply()` parallel versions of `lapply()`, `sapply()`, `apply()` (*snowfall*)

Packages

multicore parallel processing of R code on machines with multiple cores or CPUs

snow simple parallel computing in R

snowfall usability wrapper around *snow* for easier development of parallel R programs

snowFT extension of *snow* supporting fault tolerant and reproducible applications, and easy-to-use parallel programming

Rmpi interface (Wrapper) to MPI (Message-Passing Interface)

rpvm R interface to PVM (Parallel Virtual Machine)

nws provide coordination and parallel execution facilities

foreach foreach looping construct for R

doMC foreach parallel adaptor for the *multicore* package

doSNOW foreach parallel adaptor for the *snow* package

doMPI foreach parallel adaptor for the *Rmpi* package

doParallel foreach parallel adaptor for the *multicore* package

doRNG generic reproducible parallel backend for foreach Loops

GridR execute functions on remote hosts, clusters or grids

fork R functions for handling multiple processes

Generating Reports

`Sweave()` mixing text and R/S code for automatic report generation (*utils*)

R2HTML making HTML reports

R2PPT generating Microsoft PowerPoint presentations

Interface to Weka

Package *RWeka* is an R interface to Weka, and enables to use the following Weka functions in R.

Association rules:

`Apriori()`, `Tertius()`

Regression and classification:

`LinearRegression()`, `Logistic()`, `SMO()`

Lazy classifiers:

`IBk()`, `LBR()`

Meta classifiers:

`AdaBoostM1()`, `Bagging()`, `LogitBoost()`, `MultiBoostAB()`, `Stacking()`, `CostSensitiveClassifier()`

Rule classifiers:

`JRip()`, `M5Rules()`, `OneR()`, `PART()`

Regression and classification trees:

`J48()`, `LMT()`, `M5P()`, `DecisionStump()`

Clustering:

`Cobweb()`, `FarthestFirst()`, `SimpleKMeans()`, `XMeans()`, `DBScan()`

Filters:

`Normalize()`, `Discretize()`

Word stemmers:

`IteratedLovinsStemmer()`, `LovinsStemmer()`

Tokenizers:

`AlphabeticTokenizer()`, `NGramTokenizer()`, `WordTokenizer()`

Editors/GUIs

Tinn-R a free GUI for R language and environment

RStudio a free integrated development environment (IDE) for R

rattle a graphical user interface for data mining in R

Rpad workbook-style, web-based interface to R

RPMG graphical user interface (GUI) for interactive R analysis sessions

gWidgets a toolkit-independent API for building interactive GUIs

Red-R An open source visual programming GUI interface for R

R AnalyticFlow a software which enables data analysis by drawing analysis flowcharts

latticist a graphical user interface for exploratory visualisation

Other R Reference Cards

R Reference Card, by Tom Short
http://rpad.googlecode.com/svn-history/r16/Rpad_homepage/
R-refcard.pdf or
http://cran.r-project.org/doc/contrib/Short-refcard.pdf

R Reference Card, by Jonathan Baron
http://cran.r-project.org/doc/contrib/refcard.pdf

R Functions for Regression Analysis, by Vito Ricci
http://cran.r-project.org/doc/contrib/Ricci-refcard-regression.pdf

R Functions for Time Series Analysis, by Vito Ricci
http://cran.r-project.org/doc/contrib/Ricci-refcard-ts.pdf

RDataMining Website, Twitter, Groups & Package

RDataMining Website:	http://www.rdatamining.com
Twitter:	http://twitter.com/rdatamining
Group on LinkedIn:	http://group.rdatamining.com
Group on Google:	http://group2.rdatamining.com
RDataMining Package:	http://www.rdatamining.com/package http://package.rdatamining.com

Bibliography

Adler, D., Murdoch, D., 2012. rgl: 3D visualization device system (OpenGL). R package version 0.92.879.

Agrawal, R., Srikant, R., 1994. Fast algorithms for mining association rules in large databases. In: Proceedings of the 20th International Conference on Very Large Data Bases, Santiago, Chile, pp. 487–499.

Agrawal, R., Faloutsos, C., Swami, A.N., 1993. Efficient similarity search in sequence databases. In: Lomet, D. (Ed.), Proceedings of the Fourth International Conference of Foundations of Data Organization and Algorithms (FODO), Chicago, Illinois. Springer Verlag, pp. 69–84.

Alcock R.J., Manolopoulos Y., 1999. Time-Series Similarity Queries Employing a Feature-Based Approach. In Proceedings of the 7th Hellenic Conference on Informatics. Ioannina, Greece, August 27–29.

Aldrich, E., 2010. wavelets: A package of funtions for computing wavelet filters, wavelet transforms and multiresolution analyses. <http://cran.r-project.org/web/packages/wavelets/index.html>.

Breunig, M.M., Kriegel, H.-P., Ng, R.T., Sander, J., 2000. LOF: identifying density-based local outliers. In: SIGMOD '00: Proceedings of the 2000 ACM SIGMOD international conference on Management of data, ACM Press, New York, NY, USA. pp. 93–104.

Buchta, C., Hahsler, M., and with contributions from Daniel Diaz, 2012. arulesSequences: mining frequent sequences. R package version 0.2-1.

Burrus, C.S., Gopinath, R.A., Guo, H., 1998. Introduction to Wavelets and Wavelet Transforms: A Primer. Prentice-Hall, Inc.

Butts, C.T., 2010. sna: tools for social network analysis. R package version 2.2-0.

Butts, C.T., Handcock, M.S., Hunter, D.R., 2012. network: classes for relational data, Irvine, CA. R package version 1.7-1.

Chan, K.-p., Fu, A.W.-c., 1999. Efficient time series matching by wavelets. In: Internation Conference on Data Engineering (ICDE '99), Sydney.

Chan, F.K., Fu, A.W., Yu, C., 2003. Harr wavelets for efficient similarity search of time-series: with and without time warping. IEEE Transactions on Knowledge and Data Engineering 15 (3), 686–705.

Chang, J., 2011. lda: collapsed Gibbs sampling methods for topic models. R package version 1.3.1.

Cleveland, R.B., Cleveland, W.S., McRae, J.E., Terpenning, I., 1990. Stl: a seasonal-trend decomposition procedure based on loess. Journal of Official Statistics 6 (1), 3–73.

Csardi, G., Nepusz, T., 2006. The igraph software package for complex network research. InterJournal, Complex Systems, 1695.

Ester, M., Kriegel, H.-P., Sander, J., Xu, X., 1996. A density-based algorithm for discovering clusters in large spatial databases with noise. In: KDD, pp. 226–231.

Feinerer, I., 2010. tm.plugin.mail: text mining e-mail plug-in. R package version 0.0-4.

Feinerer, I., 2012. tm: text mining package. R package version 0.5-7.1.

Feinerer, I., Hornik, K., Meyer, D., 2008. Text mining infrastructure in R. Journal of Statistical Software 25 (5).

Fellows, I., 2012. wordcloud: word clouds. R package version 2.0.

Filzmoser, P., Gschwandtner, M., 2012. mvoutlier: multivariate outlier detection based on robust methods. R package version 1.9.7.

Frank, A., Asuncion, A., 2010. UCI Machine Learning Repository. School of Information and Computer Sciences, University of California, Irvine.
<http://archive.ics.uci.edu/mlurlhttp://archive.ics.uci.edu/ml>.

Gentry, J., 2012. twitteR: R based Twitter client. R package version 0.99.19.

Giorgino, T., 2009. Computing and visualizing dynamic timewarping alignments in R: the dtw package. Journal of Statistical Software 31 (7), 1–24.

Grün, B., Hornik, K., 2011. Topicmodels: an R package for fitting topic models. Journal of Statistical Software 40 (13), 1–30.

Hahsler, M., 2012. arulesNBMiner: mining NB-frequent itemsets and NB-precise rules. R package version 0.1-2.

Hahsler, M., Chelluboina, S., 2012. arulesViz: visualizing association rules and frequent itemsets. R package version 0.1-5.

Hahsler, M., Gruen, B., Hornik, K., 2005. arules—a computational environment for mining association rules and frequent item sets. Journal of Statistical Software 14 (15).

Hahsler, M., Gruen, B., Hornik, K., 2011. arules: mining association rules and frequent itemsets. R package version 1.0-8.

Han, J., Kamber, M., 2000. Data Mining: Concepts and Techniques. Morgan Kaufmann Publishers Inc., San Francisco, CA, USA.

Hand, D.J., Mannila, H., Smyth, P., 2001. Principles of Data Mining (Adaptive Computation and Machine Learning). The MIT Press.

Handcock, M.S., Hunter, D.R., Butts, C.T., Goodreau, S.M., Morris, M., 2003. statnet: Software Tools for the Statistical Modeling of Network Data, Seattle, WA. Version 2.0.

Hennig, C., 2010. fpc: flexible procedures for clustering. R package version 2.0-3.

Hornik, K., Rauch, J., Buchta, C., Feinerer, I., 2012. textcat: N-Gram based text categorization. R package version 0.1-1.

Hothorn, T., Hornik, K., Strobl, C., Zeileis, A., 2010. Party: a laboratory for recursive partytioning. <http://cran.r-project.org/web/packages/party/>.

Hothorn, T., Buehlmann, P., Kneib, T., Schmid, M., Hofner, B., 2012. mboost: model-based boosting. R package version 2.1-2.

Hu, Y., Murray, W., Shan, Y., 2011. Rlof: R parallel implementation of Local Outlier Factor (LOF). R package version 1.0.0.

Jain, A.K., Murty, M.N., Flynn, P.J., 1999. Data clustering: a review. ACM Computing Surveys 31 (3), 264–323.

Keogh, E.J., Pazzani, M.J., 1998. An enhanced representation of time series which allows fast and accurate classification, clustering and relevance feedback. In: KDD 1998, pp. 239–243.

Keogh, E.J., Pazzani, M.J., 2000. A simple dimensionality reduction technique for fast similarity search in large time series databases. In: PAKDD, pp. 122–133.

Keogh, E.J., Pazzani, M.J., 2001. Derivative dynamic time warping. In: The First SIAM International Conference on Data Mining (SDM-2001), Chicago, IL, USA.

Keogh, E., Chakrabarti, K., Pazzani, M., Mehrotra, S., 2000. Dimensionality reduction for fast similarity search in large time series databases. Knowledge and Information Systems 3 (3), 263–286.

Komsta, L., 2011. outliers: tests for outliers. R package version 0.14.

Koufakou, A., Ortiz, E.G., Georgiopoulos, M., Anagnostopoulos, G.C., Reynolds, K.M., 2007. A scalable and efficient outlier detection strategy for categorical data. In: Proceedings of the 19th IEEE International Conference on Tools with Artificial Intelligence, vol. 02, ICTAI '07, Washington, DC, USA. IEEE Computer Society, pp. 210–217.

Lang, D.T., 2012a. RCurl: general network (HTTP/FTP/...) client interface for R. R package version 1.91-1.1.

Lang, D.T., 2012b. XML: tools for parsing and generating XML within R and S-Plus. R package version 3.9-4.1.

Liaw, A., Wiener, M., 2002. Classification and regression by randomforest. R News 2 (3),18–22.

Ligges, U., Mächler, M., 2003. Scatterplot3d—an R package for visualizing multivariate data. Journal of Statistical Software 8 (11), 1–20.

Mörchen, F., 2003. Time series feature extraction for data mining using DWT and DFT. Technical Report, Department of Mathematics and Computer Science, Philipps-University Marburg.

Maechler, M., Rousseeuw, P., Struyf, A., Hubert, M., Hornik, K., 2012. cluster: cluster analysis basics and extensions. R package version 1.14.2.

R Development Core Team, 2010a. R Data Import/Export. R Foundation for Statistical Computing, Vienna, Austria. ISBN: 3-900051-10-0.

R Development Core Team, 2010b. R Language Definition. R Foundation for Statistical Computing, Vienna, Austria. ISBN: 3-900051-13-5.

R Development Core Team, 2012. R: A Language and Environment for Statistical Computing. R Foundation for Statistical Computing, Vienna, Austria. ISBN: 3-900051-07-0.

R-core, 2012. Foreign: read data stored by Minitab, S, SAS, SPSS, Stata, Systat, dBase, ... R package version 0.8-49.

Rafiei, D., Mendelzon, A.O., 1998. Efficient retrieval of similar time sequences using DFT. In: Tanaka, K., Ghandeharizadeh, S. (Eds.), FODO, pp. 249–257.

Ripley, B., from 1999 to October 2002 Michael Lapsley, 2012. RODBC: ODBC database access. R package version 1.3-5.

Sarkar, D., 2008. Lattice: Multivariate Data Visualization with R. Springer, New York. ISBN: 978-0-387-75968-5.

Tan, P.-N., Kumar, V., Srivastava, J., 2002. Selecting the right interestingness measure for association patterns. In: KDD '02: Proceedings of the Eighth ACM SIGKDD International Conference on Knowledge Discovery and Data Mining. ACM Press, New York, NY, USA, pp. 32–41.

The Institute of Statistical Mathematics, 2012. timsac: Time series analysis and control package. R package version 1.2.7.

Therneau, T.M., Atkinson, B., Ripley, B., 2010. rpart: Recursive partitioning. R package version 3.1-46.

Torgo, L., 2010. Data Mining with R-Learning with Case Studies. Chapman and Hall/CRC.

van der Loo, M., 2010. Extremevalues, an R package for outlier detection in univariate data. R package version 2.0.

Venables, W.N., Smith, D.M., R Development Core Team, 2010. An Introduction to R. R Foundation for Statistical Computing, Vienna, Austria. ISBN: 3-900051-12-7.

Vlachos, M., Lin, J., Keogh, E., Gunopulos, D., 2003. A wavelet-based anytime algorithm for k-means clustering of time series. In: Workshop on Clustering High Dimensionality Data and Its Applications, at the Third SIAM International Conference on Data Mining, San Francisco, CA, USA.

Wickham, H., 2009. ggplot2: Elegant Graphics for Data Analysis. Springer, New York.

Witten, I., Frank, E., 2005. Data Mining: Practical Machine Learning Tools and Techniques, second ed. Morgan Kaufmann, San Francisco, CA, USA.

Wu, Y.-l., Agrawal, D., Abbadi, A.E., 2000. A comparison of DFT and DWT based similarity search in time-series databases. In: Proceedings of the Ninth ACM CIKM International Conference on Informationand Knowledge Management, McLean, VA, pp. 488–495.

Wu, H.C., Luk, R.W.P., Wong, K.F., Kwok, K.L., 2008. Interpreting TF-IDF term weights as making relevance decisions. ACM Transactions on Information Systems 26 (3), 13:1–13:37.

Zaki, M.J., 2000. Scalable algorithms for association mining. IEEE Transactions on Knowledge and Data Engineering 12 (3), 372–390.

Zhao, Y., Zhang, S., 2006. Generalized dimension-reduction framework for recent-biased time series analysis. IEEE Transactions on Knowledge and Data Engineering 18 (2), 231–244.

Zhao, Y., Cao, L., Zhang, H., Zhang, C., 2009a. Data Clustering. Handbook of Research on Innovations in Database Technologies and Applications: Current and Future Trends. Information Science Reference, pp. 562–572. ISBN: 978-1-60566-242-8.

Zhao, Y., Zhang, C., Cao, L. (Eds.), 2009b. Post-Mining of Association Rules: Techniques for Effective Knowledge Extraction. Information Science Reference, Hershey, PA. ISBN: 978-1-60566-404-0.

General Index

3D surface plot, 22

APRIORI, 92
ARIMA, 78, 147
association rule, 89, 216
AVF, 73

bar chart, 15
big data, 181, 218
box plot, 17, 63

chi-square test, 165
CLARA, 53
classification, 216
clustering, 51, 55, 114, 116
confidence, 89, 92
contour plot, 22
corpus, 106
CRISP-DM, 1

data cleansing, 218
data exploration, 11, 145, 160
data imputation, 162
data mining, 1, 214
data transformation, 218
DBSCAN, 57, 70
decision tree, 27, 166, 186
density-based clustering, 57
discrete wavelet transform, 84
discretization, 157
document-term matrix,
        see term-document matrix,
        110
DTW, *see* dynamic time
        warping, 79, 82
DWT, *see* discrete wavelet
        transform, 84

dynamic time warping, 79

ECLAT, 92

forecasting, 78, 147

generalized linear model, 48
generalized linear regression, 48

heat map, 20
hierarchical clustering, 56, 80,
        82, 114
histogram, 14

IQR, 17, 143

k-means clustering, 51, 71, 116
k-medoids clustering, 53, 118
k-NN classification, 86

level plot, 21
lift, 89, 95
linear regression, 41
local outlier factor, 66
LOF, see local outlier factor, 66
logistic regression, 47

non-linear regression, 50

ODBC, 8
outlier, 58

PAM, 53, 118
parallel computing, 218
parallel coordinates, 23, 99
pie chart, 15
prediction, 216
principal component, 67

R, 2, 213
random forest, 36, 183
redundancy, 96
reference card, 213
regression, 41, 213

SAS, 6, 201
scatter plot, 18
scoring, 176
seasonal component, 76, 145,
       146
silhouette, 54, 121
snowball stemmer, 108
social network analysis, 123, 217
spatial data, 217
stemming, *see* word stemming,
       108
STL, 72
support, 89, 92

tag cloud, *see* word cloud, 113
term-document matrix, 110
text mining, 105, 217
TF-IDF, 111
time series, 72, 75
time series analysis, 213, 216
time series classification, 83
time series clustering, 78
time series decomposition, 76,
       145
time series forecasting, 78, 147
Titanic, 90
topic model, 121
topic modeling, 136
Twitter, 105, 123

word cloud, 105, 113
word stemming, 108

Package Index

arules, 92, 96, 103, 216
arulesNBMiner, 103
arulesSequences, 103
arulesViz, 99, 217
ast, 77

bigmemory, 218

cluster, 53

data.table, 218
datasets, 90
DMwR, 66
dprep, 66
dtw, 79

extremevalues, 73

filehash, 219
foreach, 218
foreign, 6
fpc, 53, 57, 59, 118

ggplot2, 24, 112
graphics, 22

igraph, 123, 124, 136

lattice, 21–24
lda, 121, 136

MASS, 23
mboost, 3
multicore, 69, 73
mvoutlier, 73

network, 136

outliers, 73

party, 27, 28, 37, 83, 166,
        182–185, 201

randomForest, 27, 36, 37, 183
RANN, 86
RCurl, 105
rgl, 20
rJava, 108
Rlof, 69, 73
rmr, 218
ROCR, 216
RODBC, 8
rpart, 27, 31, 34, 216
RWeka, 108
RWekajars, 108

scatterplot3d, 20
sfCluster, 219
sna, 136, 217
snow, 219
Snowball, 108
snowfall, 219
statnet, 136, 217, 218
stats, 77

textcat, 121
timsac, 77
tm, 105, 106, 111, 121, 217
tm.plugin.mail, 121
topicmodels, 121, 136
twitteR, 105

wavelets, 84
wordcloud, 105, 113

XML, 105

Function Index

abline(), 36, 139
aggregate(), 17
apriori(), 92
as.Date(), 138
as.PlainTextDocument(), 107
attributes(), 11
axis(), 41

barplot(), 15, 113
biplot(), 68
bmp(), 25
boxplot(), 17, 143
boxplot.stats(), 63

cforest(), 37, 184
clara(), 53
colMeans(), 142
colSums(), 142
contour(), 22
contourplot(), 22
coord_flip(), 112
cor(), 16, 164
cov(), 16
ctree(), 27, 28, 30, 31, 83, 166,
        173, 182, 185–187
cumsum(), 167, 204
cut(), 157

decomp(), 77
decompose(), 76, 146
delete.edges(), 130
delete.vertices(), 129
density(), 14
dev.off(), 25
dim(), 11
dist(), 20, 115
dtw(), 79

dtwDist(), 79
dwt(), 85

E(), 126
eclat(), 92

filled.contour(), 22
findAssocs(), 113
findFreqTerms(), 112

gc(), 196
getTransformations(), 107
glm(), 47, 48
graph.adjacency(), 124
graphics.off(), 25
grep(), 110
grey.colors(), 21
grid(), 139
gsub(), 107

hclust(), 56, 115
head(), 12
heatmap(), 20
hist(), 14

idwt(), 85
importance(), 39
interestMeasure(), 96
is.subset(), 98

jitter(), 18, 165
jpeg(), 25

kmeans(), 51, 116

levelplot(), 21
lm(), 41, 42
load(), 5

lof(), 69
lofactor(), 66, 69
lower.tri(), 98

margin(), 39
mean(), 14
median(), 14
memory.limit(), 185
memory.profile(), 185
memory.size(), 185

names(), 11
nei(), 133
neighborhood(), 134
nls(), 50

object.size(), 185, 187
odbcClose(), 8
odbcConnect(), 8

pairs(), 19
pam(), 53–55, 118
pamk(), 53–55, 118, 121
parallelplot(), 23
parcoord(), 23
pdf(), 25, 167
persp(), 22
pie(), 15
plane3d(), 45
plot(), 18
plot3d(), 20
plotcluster(), 59
png(), 25
postscript(), 25
prcomp(), 68
predict(), 27, 31, 43, 194

quantile(), 14

rainbow(), 21, 113
randomForest(), 36
range(), 14
read.csv(), 5, 137
read.ssd(), 6
read.table(), 80
read.xport(), 8

removeNumbers(), 107
removePunctuation(), 107
removeURL(), 107
removeWords(), 107
residuals(), 44
rgb(), 126
rm(), 5
rowMeans(), 142
rownames(), 111
rowSums(), 112
rpart(), 31
runif(), 60

save(), 5
scatterplot3d(), 20, 45
set.seed(), 116
sqlQuery(), 8
sqlSave(), 8
sqlUpdate(), 8
stemCompletion(), 108
stemDocument(), 107
stl(), 72, 77, 145
str(), 11
stripWhitespace(), 107
strptime(), 137
summary(), 13, 143

t(), 124
table(), 15, 39
tail(), 12
TermDocumentMatrix(), 110
tiff(), 25
tm_map(), 107, 110
ts(), 145
tsr(), 77

userTimeline(), 105

V(), 126
var(), 14
varImpPlot(), 39

with(), 18
wordcloud(), 113
write.csv(), 5

Printed and bound by CPI Group (UK) Ltd, Croydon, CR0 4YY

03/10/2024

01040418-0002